Decolonizing the Sustainable Development Goals

Reetta Toivanen • Vladislava Vladimirova
Carl-Gösta Ojala
Editors

Decolonizing the Sustainable Development Goals

Community Perspectives, Social Justice, and the Challenges of Pluralism

palgrave
macmillan

Editors
Reetta Toivanen
Department of Cultures
University of Helsinki
Helsinki, Finland

Carl-Gösta Ojala
Department of Archaeology
Ancient History and Conservation
Uppsala University
Uppsala, Sweden

Vladislava Vladimirova
Institute for Russian and Eurasian Studies/
Department of Cultural Anthropology and
Ethnology
Uppsala University
Uppsala, Sweden

ISBN 978-3-032-14262-7 ISBN 978-3-032-14263-4 (eBook)
https://doi.org/10.1007/978-3-032-14263-4

This work was supported by the Research Council of Finland's Centre of Excellence in Law, Identity and the European Narratives, and the Government of Canada's New Frontiers in Research Fund (NFRF), NFRFT-2020-00188, via Pathway T5 of the project Ărramăt: Strengthening Health and Wellbeing Through Indigenous-Led Conservation and Sustainable Relationships With Biodiversity.

This Palgrave Macmillan imprint is published by the registered company Springer Nature Switzerland AG.
The registered company address is: Gewerbestrasse 11, 6330 Cham, Switzerland

The original version of the book has been revised. A correction to this book can be found at https://doi.org/10.1007/978-3-032-14263-4_16

Acknowledgements

We are indebted to our main funders for their financial support, which also made it possible to publish this anthology with open access: the Research Council of Finland; the Centre of Excellence in Law, Identity and the European Narratives (decision number 1353311); and the Faculty of Arts at the University of Helsinki. The anthology started with the interdisciplinary workshop *Decolonizing Sustainable Development Goals* that took place at Uppsala University in September 2023. We received funding from the Uppsala Forum on Democracy, Peace, and Justice, for which we express our gratitude. The Institute for Russian and Eurasian Studies at Uppsala University co-sponsored the workshop and supported the work on the anthology in multiple ways, which we acknowledge here. We are very grateful to all participants at the workshop, and of course all of the authors who have contributed to this volume. Many colleagues from Helsinki, Uppsala, and other universities have provided invaluable help reading and commenting on drafts at various stages and deserve our gratitude. In particular, we would like to thank Ann-Mari Sätre, Leo Granberg, and Nathan Light. We are deeply grateful for the support we received from Syksy Byholm, University of Helsinki, during the editing and submission phase of this volume. Finally, we would like to mention the project "Dynamics of Circumpolar Land Use and Ethnicity (CLUE): Social Impacts of Policy and Climate Change", 2009–2013, funded by the American National Science Foundation, award number 213665, which provided the cover photo of the book.

Contents

Notes on Contributors

Gulnara Aitpaeva is the Director of the Aigine Cultural Research Center in Bishkek, Kyrgyzstan. Her research interests lie in the anthropology of sacred sites, Kyrgyz literature, and folklore. She is the co-editor of *1916: Depoliticizing and Humanizing Our Knowledge on the Uprising in Central Asia* (Bishkek, 2020), *The National Manual for Safeguarding Sacred Sites and Ritual Practices* (Bishkek, 2020) and 'Literature and Society in Central Asia', in *Cahiers D'Asie Centrale* (Paris and Bishkek, 2015).

Benedikte Brincker is Professor in Political Sociology and Head of Department at the Department of Sociology, University of Copenhagen. She is also adjunct professor at the Department of Social Science, Economics and Journalism, the University of Greenland, Ilisimatusarfik, with research interests in state-nation relations and nationalism; the sociology of education; public administration.

Andrei Danilov represents the Sámi people in the Kola Peninsula. He has been a member of the Organization of Sámi of Murmansk Region (OOSMO) and a founder of the Sámi National Cultural Autonomy in the town of Monchegorsk. In 2008, he was elected a member of the Kola Sámi Parliament. Since 2011, he has been the Sámi Heritage and Development Fund director. From 2013 to 2017, he was a vice president of the Sámi Council. With Lipka O. N., Sovkina V. V., and Sulyandziga R. V. L, he co-authored the books *Arctic Nature and Indigenous Populations under the Impact of Climate Change and Industrial Development: Murmansk Oblast* (2021) and *Lovozero, Murmansk Oblast. Climate Change Adaptation Strategy for Villagers* (2022).

Elle Eriksson has an MSc in Forestry. She is a reindeer herding member of Váhtjer Sámi community in Sápmi, northern Sweden.

Tobias Haller is a professor at the Institute of Social Anthropology, University of Bern, Switzerland. He has specialized in ecological and economic anthropology and done research in different African countries on common property institutions as well as colonial and post-colonial land and green grabbing in the context of development projects and conservation. He is also doing research on local institution building processes and local environmental perceptions. Together with Ikhom Soliev, he is the co-coordinator of the Regional Chapter of the International Association for the Study of the Commons (IASC) for Europe and CIS.

Marjo Lindroth is a senior researcher in the Arctic Centre at the University of Lapland, Finland. Her research critically investigates the intersections between power, rights and indigeneity in international politics. Her particular research interests at present are hope, resilience and contemporary colonialism. Her latest books include, among others, *The Colonial Politics of Hope: Critical Junctures of Indigenous-State Relations* (co-authored with Heidi Sinevaara-Niskanen, Routledge, 2022).

Elisa Maria López is a cultural anthropologist and postdoctoral researcher at KTH Royal Institute of Technology School of Architecture, Stockholm. Her scholarship focuses on environmental relations and the politics of space in far northern Nordic industrial cities.

Mark Nuttall is a professor and Henry Marshall Tory Chair in the Department of Anthropology at the University of Alberta in Edmonton, Canada. He is also an adjunct professor at Ilisimatusarfik/University of Greenland and Greenland Climate Research Centre in Nuuk. He has published extensively on climate change and the Arctic, i.e., in 2017 the award-winning book *Climate, Society and Subsurface Politics in Greenland: Under the Great Ice* (London: Routledge).

Johanna Ohlsson is Senior Lecturer in Human Rights and Democracy at University College Stockholm and affiliated researcher to the Institute for Russian and Eurasian Studies and Centre for Multidisciplinary Research on Religion and Society at Uppsala University.

May-Britt Öhman is researcher at the Centre for Multidisciplinary Studies on Racism, CEMFOR, at Uppsala University, Sweden. She is Associate Professor of Environmental History and holds a PhD in History of Technology. Öhman is a Lule and Forest Sámi of the Lule River Valley, with roots also in Torne River Valley and Roslagen. She is currently a board member of the Sámi journal *Samefolket*, and also serving as expert within the Swedish government–appointed Committee on Reindeer Lands (*Renmarkskommittén*), on behalf of the National association Sámi Ätnam.

Carl-Gösta Ojala is Senior Lecturer in Archaeology at Uppsala University, Sweden, and an associate professor at the Global Institution for Collaborative Research and Education, Hokkaido University, Japan. His main research interests include archaeology and heritage in Northern Fennoscandia and northern Russia, Sámi archaeology and heritage management, critical heritage studies, politics of the past, debates on Indigenous rights and decolonization, and the interrelations of heritage and sustainable development.

Tuija von der Pütten is a doctoral researcher at University of Helsinki and a member of the Helsinki Institute for Sustainability Science (HELSUS). She is currently working on her doctoral research on corporate climate responsibility from the perspective of EU and international law. Her research interests include international climate law, corporate responsibility/sustainability, EU corporate sustainability framework, clean investments and investment treaties.

Heidi Sinevaara-Niskanen a senior lecturer and feminist scholar at the University of Lapland, Finland. She holds a Title of Docent in World Politics at the University of Helsinki. Her critical research addresses the problematics of development, power and social sustainability, with particular reference to Arctic politics. Her latest books include, among others, *The Colonial Politics of Hope: Critical Junctures of Indigenous-State Relations* (co-authored with Marjo Lindroth Routledge, 2022).

Tracey Skillington is a senior lecturer in the Department of Sociology & Criminology and Director of the BA (Sociology) as well as an affiliate of the Environmental Research Institute, University College Cork. Her publications include *Climate Justice & Human Rights* (2017, Palgrave), *Climate Change & Intergenerational Justice* (2019, Routledge).

Ilkhom Soliev is a professor and Head of the Department for Environmental Sociology at the Institute for Sociology of Martin Luther University Halle-Wittenberg (MLU), Germany. He is also the Director of the Social Learning and Environmental Governance Lab at the Centre for Interdisciplinary Regional Studies of MLU. His research focus is on environmental social sciences, resource governance, institutional economics, and societal transformations in relation to nature, biodiversity, and commons. Together with Tobias Haller he is the co-coordinator of the Regional Chapter of the International Association for the Study of the Commons (IASC) for Europe and CIS.

Reetta Toivanen is Full Professor of Sustainability Science (Indigenous Sustainabilities) at the Helsinki Institute for Sustainability Science (HELSUS) and Department of Cultures at the Faculty of Arts at the University of Helsinki, Finland. During 2023–2025 she is a member of the Finnish Government's Expert Panel on Sustainability.

Anna Niia Varfolomeeva is a postdoctoral researcher in the Biodiverse Anthropocenes program, at the University of Oulu, Finland. Her postdoctoral project centers on the influence of wind energy and mining on human-landscape relations in the Finnish and Russian Arctic. Previously, Varfolomeeva worked at the University of Helsinki and the University of Tyumen. She has published on Indigeneity construction in resource-rich regions and on the symbolism of stone extraction.

Vladislava Vladimirova is Associate Professor of Cultural Anthropology and Senior Lecturer in Eurasian Studies at the Institute for Russian and Eurasian Studies (IRES) and the Department of Cultural Anthropology and Ethnology at Uppsala University. She is involved in long-term research on Indigenous rights, environmental governance and justice, gender, moral economy, and conservation of environmental and cultural heritage in the Arctic.

List of Figures

List of Tables

CHAPTER 1

Decolonizing the Sustainable Development Goals: Introduction

Reetta Toivanen, Vladislava Vladimirova, and Carl-Gösta Ojala

This book presents innovative research critically examining the effects of the implementation of the Sustainable Development Goals (SDGs) on Indigenous and local communities in the Arctic, and globally. As climate change has growing effects on the world, there is a need for in-depth understanding of its future transformational impact on numerous areas of public law and policy, such as those aimed at implementing the SDGs (Cambou, 2019). The tensions between actions addressing the effects of climate change and legal protection of human rights is an especially vital topic (Pál & Brain, 2018). These tensions are highly visible in more extreme or fragile environments, such as arid areas and the Arctic. The Arctic is a vast part of the planet, once seen as a wild, empty, and remote area, which is now increasingly understood as a strategic

R. Toivanen (✉)
Department of Cultures, University of Helsinki, Helsinki, Finland
e-mail: reetta.toivanen@helsinki.fi

V. Vladimirova
Institute for Russian and Eurasian Studies/Department of Cultural Anthropology and Ethnology, Uppsala University, Uppsala, Sweden
e-mail: vladislava.vladimirova@ires.uu.se

C.-G. Ojala
Department of Archaeology, Ancient History and Conservation, Uppsala University, Uppsala, Sweden
e-mail: carl-gosta.ojala@arkeologi.uu.se

R. Toivanen et al. (eds.), *Decolonizing the Sustainable Development Goals*,
https://doi.org/10.1007/978-3-032-14263-4_1

region, with expanding contests over resources, territories, and military empowerment (e.g., Nuttall, this volume). Climate change is tangible in the Arctic where melting permafrost and sea ice, on the one hand, and open access to new mineral resources and trading routes, on the other hand, increase the vulnerability of local human and non-human populations and industrial and military infrastructures (Toivanen, 2019; Vladimirova, 2015). The Arctic is simultaneously a region of important resource extraction, transport infrastructure, military bases, and large urban centers, and home to over 40 Indigenous peoples whose living environment, subsistence, and cultural and linguistic heritage are entitled to preservation (Nuttall, 1998).

In recent decades, much attention in national and international politics, civil activism, and academic fora has been devoted to climate change impacts and their mitigation in the Arctic. Many initiatives and documents envisage a better understanding of climate change and how its effects intersect with other biophysical, social, political, and economic processes. The SDGs are among the globally most recognized and applied directives addressing climate change reduction and mitigation.[1] The United Nations' Agenda 2030 is a plan of action for people, the planet and prosperity. In order to implement it, governments and other stakeholders met in 2015 in New York and formulated 17 SDGs. Agenda 2030 is a vigorous statement of policy objectives, often compared with the United Nations Declaration of Human Rights, defining clear moral principles and guidelines. Although Agenda 2030 has gained a strong footing in the design of government programs and even research funding, it nevertheless constitutes a human-made political compromise. Its critics note that "Agenda 2030 does not challenge the positions of powerful actors such as big countries, international financial institutions, transnational corporations and even international NGOs that have produced and reproduced inequalities in income, wealth and power at national and global levels, amplifying the very problems the SDGs are trying to solve" (Esquivel, 2016; Toivanen, this volume). The SDGs combine more recent ambitions and visions about limiting climate change through sustainable use of resources, with older frameworks promoting equality and wellbeing in economic development, and with human rights through democracy and civil society initiatives. The SDGs also offer a mechanism for evaluating progress in implementing particular goals within and between different groups and nations. Analysts have been both positive and critical to the quantitative measurement of the implementation of the SDGs, and the comparisons and ranking of states. Quantitative measurements help capture progress and regress in achieving goals, and comparison can help identify successful policies and processes, and their transfer among nations (spillover).[2] At the same time, the SDG targets whose implementation is being measured are narrowly formulated and allow for inaccuracies, exaggerated achievements, and omissions in capturing negative trends (Dalby et al., 2019).

[1] https://sdgs.un.org/goals.
[2] https://dashboards.sdgindex.org/rankings/spillovers.

Chapters in this volume also bring into focus different instances of SDG formulation and targets that miss to address sources of climate change and societal inequalities. SDG inscription in developmental agendas prioritizes areas and paths of development and equality that reproduce the visions, values, experiences, and interests of certain regions and societal groups, for example, the Western urban middle class. These groups seem to have modeled the implicit standards and aspirations that the SDGs promote in the world. As authors in this volume argue, the SDGs ignore and thus undermine efficient alternative models of sustainable nature governance such as Indigenous and local environmental knowledge (Eriksson & Öhman, this volume; Brincker, this volume), common property regimes (Haller & Soliev, this volume), and women's equality through traditional societal institutions of status and respect (Aitpaeva & Vladimirova, this volume). Furthermore, the implementation of the SDGs often lacks historical perspectives, in particular consideration of colonial and imperial histories and legacies, as well as Indigenous histories (Ojala, this volume; López, this volume).

As critics of the SDGs argue, the emphases and dominant measurements of green growth and alternative sources of green energy have led to disregard of other forms of nature use that are in conflict with or dismantled by the green economy (e.g., Skillington, this volume). Chapters in the anthology show how wind parks in Northern Europe and mining for minerals deemed essential for the green transition threaten Indigenous and local economies and cultures (e.g., von der Pütten, this volume; Varfolomeeva, this volume). Protecting the rights of Indigenous communities and people is not explicitly mentioned in the SDGs, and they do not align well with and, at times, even contradict local communities' interests (Varfolomeeva, 2017; Ohlsson, this volume). While national legislation and bureaucratic regulations on community participation, the application of Free, Prior, and Informed Consent (FPIC), and other forms of deliberative democracy vary, Indigenous residents are often refused equal partnership in decision-making regarding SDG implementation in their region (e.g., Danilov, this volume; Eriksson and Öhman, this volume). This book addresses the exclusion of Indigenous communities and explores the practices of national governments, regional authorities, and business actors that contribute to exclusion. The authors also examine possible ways to improve partnerships with Indigenous and local communities, including better integration and alignment between Indigenous perspectives and formulations and implementations provided for in the SDGs.

Critical social scientists warn that Agenda 2030 and the SDGs set to achieve it have fundamentally failed to engage with alternative ways of perceiving development. Generalized notions of sustainability bear the risk of overshadowing and suppressing those whose perspectives differ from mainstream discourses, including Indigenous communities. Instead of addressing global diversity and the inequality of political participation, Agenda 2030 primarily deals with Western-centered science, knowledge, and rationality when analyzing needs, setting priorities, or finding solutions (Struckmann, 2018). These tensions and inequalities in the implementation of the SDGs are particularly visible in the Arctic region and in states where agriculture and small-scale household economies are widespread.

The Arctic is a place of great diversity, both environmentally and socially. Climate change has caused warming in the Arctic at more than seven times the global average which drastically impacts Northern communities but also has impacts all over the globe. The manifold consequences of environmental change include thawing permafrost, shrinking sea ice cover, biodiversity loss, coastal erosion and sea level rise, shifts in animal migrations, and ecosystem disturbances. These processes intersect in complex and unpredictable ways with economic and political events, such as the Russian war against Ukraine, which has led to further militarization in the global Arctic, wider military environmental pollution, and increased extraction of mineral resources used in military industry (Vladimirova, 2024; Danilov, this volume). The current energy crisis and economic inflation have also lowered many requirements for sustainable food production and transport, which slows the implementation of the SDGs and exacerbate the impacts of climate change. Challenges to Arctic communities are particularly acute due to the high price of food production and transport services, which increase the vulnerability of Indigenous peoples. Hindrances to Indigenous economies, such as reindeer herding, fishing, and hunting, imposed by competing mining industries, militarization, and tourism, add to deteriorating environmental conditions, including later starts of winter and spring. The higher rates in mortality of reindeer and other species not only cause disturbances in and threats to Indigenous subsistence, but violate important Indigenous cultural, cosmological, and ontological orders (Nuttall, 1998). Increasing political and economic interest in Arctic landscapes and seascapes contributes to potential conflicts between Indigenous communities and industrial activities (Gladun et al., 2007). Sustainable development, including economic, social, and environmental aspects, is hence crucial for Arctic communities and has been an official goal of the Arctic Council since its establishment by the Ottawa Declaration in 1996. The specific circumstances of Arctic livelihoods complicate the application of the SDGs as they do not fully take into account the region's characteristics. Arctic conditions include demographic challenges caused by out-migration and increasing urbanization, Indigenous self-determination and rights, the diversity of Arctic economies, as well as heightened geopolitical tensions. Considering these conditions and the potential conflicts arising between various stakeholders, there is a need to evaluate the role and implementation of the SDGs in the Arctic. In order to better grasp the role of specific conditions and how they should be considered in the implementation of the SDGs, we need also comparative perspectives. Therefore, a number of chapters in this anthology provide insights and knowledge about the impact of the SDGs on Indigenous and local groups in other parts of the world, such as Central Asia, Africa, and Mesoamerica.

Anthropologists have contributed to the research on climate change and sustainable development through in-depth analysis of local experience, emic conceptions, and knowledge of nature in a longer durée (Baer & Singer, 2018; Crate & Nuttall, 2016; Dove, 2013). In the last decade, anthropology of climate change has been at the forefront of the push for knowledge co-production with Indigenous and local stakeholders (Lavrillier & Gabyshev, 2017). Research has engaged with institutions that have hegemonic power over the environment and resources, and with Indigenous peoples and knowledges. Nonetheless, questions remain: Have anthropologists sufficiently problematized the power dynamics of environmental knowledge production itself? Has anthropological research worked adequately at producing synergetic knowledge in cooperation with other disciplines?

Authors from different disciplinary fields met at the workshop "Decolonizing Sustainable Development Goals" at Uppsala University, Sweden, in September 2023 to cross disciplinary borders and, in deliberation, complement anthropological debates. Papers that were presented at the workshop evolved and developed as a result of cross-disciplinary discussions. We took a starting point in critical thinkers of decolonial theory in a search to de-naturalize environmental knowledge and developmental epistemologies foregrounding the SDGs. We invited speakers and authors with diverse disciplinary, ethnic, and cultural backgrounds and experiences. Indigenous scholars and representatives of communities were invited to share their stories, critical insights, and analyses as essential parts of joint knowledge production and pluriverse and decolonial thinking (Reiter, 2018). Due to the persistence of myriad more or less perceptible ways of stigmatization and discrimination practices in academia and society, we let authors in the book choose how they want to present themselves and their knowledge and how they want to disclose their identities and relate their research to them. In this way, the editors do not single out Indigenous scholars and their chapters here. Furthermore, following the call of anthropologists and feminist, decolonial, and critical thinkers, we gather in this volume contextually situated knowledges and analytical approaches from different parts of the world that evaluate the SDGs, paying attention to multiple affected voices, narratives, uncertainties, injustices, and contestations, grounded in subjective everyday life experiences (Mignolo & Walsh, 2018). We hold that the SDGs and documents like them should be more democratic and inclusive in order to benefit from knowledge co-production, where a plurality of historical and cultural experiences, values, and human and nature's rights are reflected. In the world we live in, skepticism and pessimism about shared human values, such as humanism, mutual care, and compassion, is growing, as reflected in many ways in contemporary social and political life in different parts of the world. We suggest that by choosing the decolonial option and explicitly addressing pluriverse knowledge production, the SDGs have the potential to show the way to an important societal turn.

As the world experiences growing effects of climate change, there is a need for further ethnographic research examining the future transformational impact on numerous areas of public law and policy, and the economy. Ethnography addressing the effects of climate change through measures ensuring sustainability and protecting human rights and social justice is especially vital. Ethnography is powerful in revealing conflicts between sustainability, on the one hand, and social justice and human rights, on the other. Anthropologically led research can help understand how Indigenous residents are neglected in decision-making regarding the UN Agenda for 2030 and its implementation through the SDGs. Anthropology promotes the co-production of knowledge and community-based management of the environment in interdisciplinary research programs and in political and economic decision-making. This volume shows the huge potential of anthropology for increasing positive impacts from the SDGs through increased participation and consideration of Indigenous and local peoples. In addition, it features papers from other disciplines that generate productive synergies with anthropological knowledge. For example, we include historical perspectives in order to examine the deep historical roots of contemporary conflicts over environmental management and land use, and in order to contextualize and challenge present notions of sustainability and development (Ojala & Nordin, 2015; Ojala, this volume).

Outline of the Volume

Following this introductory chapter, Chap. 2 of the volume by Carl-Gösta Ojala highlights the need for historical perspectives in discussing sustainable development and Indigenous rights, with special focus on Sápmi, the main settlement area of the Indigenous Sámi people in northernmost Europe. Contemporary conflicts over land, resources, and Sámi heritage stem from colonial histories and an enduring "extractive mindset." Archaeology and material culture studies offer insights beyond written records, shedding light on resource exploitation, the collecting of artefacts, and human ancestral remains from the seventeenth century until today. The chapter explores how these perspectives can inform debates on decolonization, Sámi rights, and sustainability.

Chapter 3 written by Mark Nuttall examines how northern Greenland, Canada's Arctic Archipelago, and the Arctic Ocean are framed as spaces demanding urgent conservation action. While well intentioned, scientific and environmental interventions often marginalize Inuit communities by prioritizing external assessments of wildlife and ecosystems. High Arctic coastal communities also navigate economic and geopolitical narratives of resource potentials and security threats. Focusing on ice and Arctic marine mammals, the chapter explores how Inuit challenge dominant conservation models and assert their role in environmental governance. Case studies on Pikialasorsuaq, marine protected areas, and the Central Arctic Ocean fisheries agreement highlight Inuit-led initiatives for co-management and Indigenous Protected Areas, addressing SDGs 13 and 14.

Chapter 4, by Reetta Toivanen examines the tensions between climate change mitigation and Indigenous rights in the Arctic, where environmental transformation is occurring at an accelerated rate. It critically assesses the applicability of the SDGs in Arctic contexts, highlighting concerns that Agenda 2030 fails to engage with alternative development perspectives. Through synergetic analysis of SDG indicators and ethnographic research, the chapter argues that the SDGs risk perpetuating colonial logics, positioning Indigenous homelands as resource frontiers for the majority population rather than recognizing their rights and sustainable livelihoods.

Chapter 5, by Tuija von der Pütten, examines human rights violations linked to wind energy investments on Indigenous lands and the role of corporate sustainability in addressing these issues. From an international law perspective, it explores the challenges of clean energy projects developed without free, prior, and informed consent, focusing on the Fosen wind farm in Norway, the Lake Turkana wind farm in Kenya, and the Gunaa Sicarú project in Mexico. While wind energy is crucial for climate goals, such projects often undermine Indigenous rights. The chapter highlights the need for stronger corporate responsibility, mandatory human rights due diligence, and legal accountability to ensure truly sustainable renewable energy development.

Chapter 6, by Johanna Ohlsson, examines the relationship between sustainability and Indigenous rights through a decolonial lens, critically assessing human rights language in the UN SDGs. While sustainability and human rights are generally seen as interconnected, tensions emerge in SDG discourses. A content analysis of SDG reports (2016–2023) reveals that Indigenous rights are rarely addressed, highlighting the need for greater integration. The chapter also explores the colonial dimensions of sustainability efforts and the crucial role of language in shaping policy and discourse.

Chapter 7, by Elisa López, critically examines the "New Norrland" vision, which frames northern Sweden as central to the national "green transition" in industry, energy, and resource extraction. Focusing on Kiruna (Giron), a key site for green industrial expansion and a Sámi homeland, it explores the tensions between sustainability goals and Sámi rights, particularly in reindeer herding areas affected by mining activities. Through anthropological fieldwork (2022–2023) and analysis of policy and corporate materials, the chapter investigates how state and corporate actors shape this transition and how Sámi communities and local planners respond to its challenges.

Chapter 8, by Benedikte Brincker, applies Karl Weick's concept of sensemaking to the Inuit of East Greenland, exploring how Indigenous interpretations of nature provide insights into de-growth societies. Drawing on Lauren Berlant's cultural critique, it identifies three infrastructures shaping Inuit polar bear hunting practices: governance, rewards, and insurance. The chapter examines these dimensions and assesses their contributions to understanding de-growth.

Elle Eriksson and May-Britt Öhman have co-authored Chap. 9. In their chapter they point out that the EU aims to be climate-neutral by 2050, with the legally binding European Green Deal aligning with SDG 7's goal of ensuring sustainable energy access. However, wind power projects have repeatedly violated Sámi rights, as they demonstrate. Their chapter examines the court case of Hällberget in the Váhtjer Sámi community, in northern Sweden, providing historical context on Sámi reindeer herding and land rights. The authors critique SDG 7 from a decolonial perspective, situating Sámi struggles within broader Indigenous and international contexts.

In Chap. 10, Heidi Sinevaara-Niskanen and Marjo Lindroth examine the political power and inequalities embedded in the aspirations and hopes reflected in the SDGs. While the SDGs have received attention, the political implications of these hopes have been overlooked. Drawing on the politics of hope, they analyze competing futures within the SDGs and question whose hopes are prioritized. Focusing on land and resource debates in northern Finland, Sweden, and Norway, they highlight Sámi peoples' divergent hopes, which challenge dominant Western sustainability visions. Decolonizing the SDGs requires reevaluating the assumptions shaping sustainability goals.

Recent climate crises highlight the ways ecological issues are intertwined with imperialism, colonialism, and marginalization. Chapter 11, authored by Tracey Skillington, examines how wildfire devastation in the Arctic reveals the reinforcement of imperial legacies through sustainable development discourse and policy. It also explores how disregard for Indigenous Arctic communities' lived experiences and knowledge undermines inclusive collective learning. The chapter then considers a deliberative process aiming to address these issues and explore wrongdoing from more localized perspectives on sustainable futures.

In Chap. 12, Gulnara Aitpaeva and Vladislava Vladimirova discuss recent feminist and Indigenous critiques of SDG 5 (which aims to "achieve gender equality and empower all women"). These critiques highlight how this goal relies on a development framework that measures women's well-being through modern social and economic progress while overlooking local and Indigenous knowledge, societal values, and culture. Many Indigenous communities view the SDGs as an external imposition contributing to colonial appropriations. Using ethnographic data from Kyrgyzstan and the Russian Arctic, this chapter offers a critical perspective on SDG 5, grounded in women's voices, experiences, and knowledge.

Renewable energy, particularly wind power, is central to Finland's climate goals, with plans to reach 51% renewable energy by 2030. While many support wind energy, opposition arises when wind power stations are proposed near homes. Chapter 13, by Anna Niia Varfolomeeva, examines local reactions to wind energy infrastructure in northern Finland, focusing on human–landscape relations. It engages with concepts like "energy democracy" and "Aeolian extractivism" to explore the broader implications of wind power, highlighting how it reinforces regional inequalities and impacts local connections to forests and animals.

In Chap. 14, Tobias Haller and Ilkhom Soliev critically reassess SDGs which attribute most of the agency for the implementation of sustainability to national governments and private capital. They ask the question of how communities in the Global South, who collectively own and manage resources, are envisioned within the green transition. Drawing on Elinor Ostrom's work on institutions that govern common pool resources, and on empirical examples, the authors provide persuasive arguments that the commons are of great relevance to sustainability and community resilience globally. The lack of recognition and explicit provisions about the commons in the SDGs poses the danger of further expansion of land privatization and commons grabbing by corporations that threaten sustainability, nature preservation, and communities.

In an epilogue to the collection, Andrei Danilov, a respected Sámi politician and activist, who has lived for some years as a political refugee in Norway, shares his tragic experiences of the expanding extractive industry, corruption, and disregard for human rights and for nature in Russia. He tells a heartbreaking story of state appropriation of nature and Indigenous culture in Sámi communities on the Kola Peninsula, resulting from projects aimed at militarization and at enriching elites.

With this wide range of case studies and in-depth discussions, this volume critically examines the tensions between addressing climate change and upholding Indigenous peoples' rights and local perspectives in the implementation of the SDGs. It highlights the often-overlooked power dynamics in environmental knowledge production, a topic that demands deeper and more thorough exploration. Focusing on Indigenous peoples and local communities in the Arctic, with particular attention to those living in the Nordic countries, the volume provides new insights into the challenges and opportunities of integrating local and Indigenous knowledge within the framework of the SDGs. We hope that this volume will contribute to the critical evaluation and discussion of the decolonization of the field of sustainable development, and to the acknowledgment and inclusion of the unique environmental knowledge and important historical experiences of Arctic Indigenous and local communities in future contextualization and implementation of the SDGs, to the benefit of our shared world.

References

Baer, H. A., & Singer, M. (2018). *The anthropology of climate change: An integrated critical perspective*. Routledge.

Cambou, D. (2019). The UNDRIP and the legal significance of the right of indigenous people to self-determination: A human rights approach with a multidimensional perspective. *International Journal of Human Rights, 23*(1–2), 34–50.

Crate, S. A., & Nuttall, M. (2016). *Anthropology and climate change: From encounters to actions*. Routledge.

Dalby, S., Horton, S., & Mahon, R. (2019). Global governance challenges in achieving the Sustainable Development Goals. In S. Dalby, S. Horton, R. Mahon & D. Thomaz (Eds.), *Achieving the Sustainable Development Goals: Global governance challenges* (pp. 1–18). Routledge.

Dove, M. R. (2013). *The anthropology of climate change: A historical reader*. John Wiley & Sons.

Esquivel, V. (2016). Power and the sustainable development goals: A feminist analysis. *Gender and Development, 24*(1), 9–23.

Gladun, E., et al. (2007). *Guarantees of rights and interests of indigenous peoples in the Russian north*. Tyumen State University.

Lavrillier, A., & Gabyshev, S. (2017). *An Arctic indigenous knowledge system of landscape, climate, and human interactions: Evenki reindeer herders and hunters*. Kulturstiftung Sibirien.

Mignolo, W. D., & Walsh, C. E. (2018). *On decoloniality: Concepts, analytics, praxis*. Duke University Press.

Nuttall, M. (1998). *Protecting the Arctic: Indigenous peoples and cultural survival*. The Gordon and Breach Publishing Group.

Ojala, C.-G., & Nordin, J. M. (2015). Mining Sápmi. Colonial histories, Sámi archaeology, and the exploitation of natural resources in northern Sweden. *Arctic Anthropology, 52*(2), 6–21.

Pál, V., & Brain, S. (2018). *Environmental politics and policy under authoritarian regimes: Myth, propaganda, reality*. Routledge.

Reiter, B. (2018). *Constructing the pluriverse: The geopolitics of knowledge*. Duke University Press.

Struckmann, C. (2018). A postcolonial feminist critique of the 2030 agenda for sustainable development: A south African application. *Agenda, 32*(1), 12–24.

Toivanen, R. (2019). European fantasy of the Arctic region and the rise of indigenous Sámi voices in the global arena. In N. Sellheim, Y. V. Zalka, & I. Kelman (Eds.), *Arctic triumph: Northern innovation and persistence* (pp. 23–40). Springer Polar Sciences.

Varfolomeeva, A. (2017). Remaking stone and community: Indigenous Vepses' engagement in mining in northwestern Russia. *Polar Geography, 41*(1), 26–38.

Vladimirova, V. (2015). The north is getting hot. The growing significance of contests over nature in circumpolar Russia. *Kungliga Humanistiska Vetenskaps-Samfundet i Uppsalas Årsbok*, 191–208.

Vladimirova, V. (2024). Continuous militarization as a mode of governance of indigenous people in the Russian Arctic. *Politics and Governance, 12*, 1–15.

CHAPTER 2

Colonialism and Sustainable Development in Sápmi: Historical and Archaeological Perspectives

Carl-Gösta Ojala

Introduction

Recent industrial developments in Sápmi—the core settlement area of the Indigenous Sámi people in northern Norway, Sweden, Finland and the Kola Peninsula in Russia—often framed within the so-called green transition, with plans for new mines, enormous windfarms and associated infrastructure, add to earlier large-scale exploitation of natural resources, including the damming of rivers and lakes, forestry, and mining. The recent expansion of extractive industries in the Sámi areas, which threatens traditional land use, reindeer grazing land and migration routes, as well as the local environment, has been met by opposition and protests from Sámi groups and environmental groups (Ojala & Nordin, 2015; Eriksson & Öhman, this volume). In these protests, history becomes important and protesters are talking about not only a colonial *past*, but also about a colonial *present*, demanding a decolonization of state and industrial policies and practices, in order to protect Indigenous land rights and work for future sustainable development.

Parallel with debates on Sámi land rights, issues of cultural rights have been increasingly discussed in recent years in Sápmi. Sámi groups have demanded

C.-G. Ojala (✉)
Department of Archaeology, Ancient History and Conservation, Uppsala University, Uppsala, Sweden
e-mail: carl-gosta.ojala@arkeologi.uu.se

R. Toivanen et al. (eds.), *Decolonizing the Sustainable Development Goals*,
https://doi.org/10.1007/978-3-032-14263-4_2

greater self-determination in heritage management, and control over colonial collections of Sámi material culture, as well as Sámi ancestral human remains.

Many of the present-day conflicts over land and cultural rights in Sápmi, including conflicts over the exploitation of natural resources and the management of Sámi heritage, have deep historical roots. In order to understand these current debates and conflicts, historical perspectives are needed.

In this chapter, I explore the historical development of an "extractive mindset" in Sápmi, which has been an important part of colonial ideologies and practices. Furthermore, I argue that material sources can play an important role in the historical explorations. There is a potential of archaeology and material culture studies to throw light on aspects that are not reflected in the written sources and to provide new perspectives. In the chapter, I discuss the interrelations of the exploitation of natural resources and the collecting of material culture and human remains in Sápmi, as part of colonial networks stretching from the early seventeenth century until today. How can archaeological and historical perspectives contribute to the understanding of current debates on decolonization, Sámi Indigenous land and cultural rights, and sustainable development?

Colonial history and heritage in Sápmi has been a contested and controversial issue in the Nordic countries. For many years, there was a general reluctance to acknowledge colonial histories and relations and a political and cultural unwillingness to recognize the Nordic states as colonial powers in Sápmi, connected with views of Nordic exceptionalism (Ojala, 2023). Although recent years have witnessed an increasing amount of research investigating colonial histories and legacies, there is still a great need to critically examine Nordic colonialism in Sápmi and the field of tension between past and present.

Histories of Extractivism in Sápmi

The early modern Scandinavian colonial expansion towards the north aimed to exert control over taxation, trade routes, and natural resources. This expansion was undertaken as part of an ideology of Christian conversion of the perceived heathen Sámi population (Fredriksson et al., 2019; Ojala, 2020). There was a competition for the Sámi areas between the Kingdoms of Sweden-Finland and Denmark-Norway, and the Russian state, which also formed part of the ideological background to the colonial expansion

From the early seventeenth century, church and market sites were established in inland Sápmi, in northern Sweden, in places such as Lycksele/Lïkssjuo, Arvidsjaur/Árviesjávrrie, Jokkmokk/Jåhkåmåhkke, Jukkasjärvi/Čohkkiras and Enontekiö/Markkina/Márkan. The local Sámi populations were required to be present at these sites during certain periods of the year, in order to trade, pay taxes, and take part in judicial and religious matters. These places functioned as nodes in the colonial landscapes, as centres for state and church authorities, where different groups interacted and ideas and material culture were negotiated, also becoming important sites in the Sámi landscapes (Bergling, 1964; Ojala & Nordin, 2019; Ojala, 2020).

Another important driving force behind the early modern colonial expansion towards the north was the hope to find and exploit valuable mineral deposits, such as silver and copper, in the Sámi areas. In the 1630s, silver ore was discovered at Nasafjäll/Násavárre in the high mountains, near Arjeplog/Árjepluovve in northern Sweden, which led to the establishment of a silver mine at Nasafjäll and a silver metal works at Silbojokk/Silbbajåhkå (Ojala & Nordin, 2015). During parts of the seventeenth century, local Sámi people were forced to work for the silver mine and works, in particular with transportation which had to take place with reindeer and sledges. The silver mine and works also functioned as centres for missionary activities, forced conversion, and the destruction and collecting of sacred Sámi objects. Nasafjäll represents the start of the exploitation of mineral resources in the Sámi areas in Sweden, which has continued until the present day and has been of paramount importance for the national economy of Sweden.

Sápmi was constructed as a space for extraction already in the seventeenth century, with high hopes for large revenues from the exploitation of natural resources. This vision of Sápmi, with roots in the seventeenth century and continuity until the present day, has been activated in the imaginaries of the so-called green transition in Sweden in recent years (which has entailed many new industrial investments and developments in the Sámi areas, as discussed by López, in this volume).

Concurrently with the state, Church and industrial colonization of Sápmi, a practice of colonial collecting also developed during the early modern period. Sámi material objects, especially sacred objects, were highly sought after by collectors and scholars in different countries (Nordin & Ojala, 2019; Snickare, 2022). The sacred Sámi drums—which were an important part of Sámi Indigenous religion in the seventeenth and eighteenth centuries—were particularly coveted, as exotic, mystic and powerful objects. In the campaigns against traditional Sámi religious beliefs and practices, the use of the drums was forbidden, with threat of serious punishment, including the death penalty. Many Sámi drums were confiscated in different places in Sápmi—a large number of drums were destroyed, but some were also collected and ended up in various collections in the Scandinavian countries and elsewhere.

During the nineteenth century—as part of the developing physical anthropological, craniological and racial anthropological research—a practice of collecting ancestral human remains in Sápmi emerged. National and international scholars, explorers and travellers, but also local officials, doctors and clergymen, took part in the plundering and theft of human remains. There are many examples of this, which took place in different parts of Sápmi, in Norway, Sweden, Finland and the Kola Peninsula in Russia, during the nineteenth and early twentieth centuries. Often it was the collectors and scholars themselves who wrote about how they struggled to find and take human remains, through grave plundering or by stealing the remains of deceased persons, without any respect or understanding for relatives' and local peoples' wishes and emotions. One example is Lars Levi Læstadius (1800–1861), vicar in Karesuando and later Pajala parishes in northernmost Sweden, and very well known as the founder of the revivalist Læstadian religious movement, who acted as a guide for visiting scholars and participated in repeated grave plundering at the cemetery in Enontekis/Markkina, which was one of the market and church sites established by the Swedish Crown in the early seventeenth century, mentioned above (see further Ojala, 2016, 2022b).

There was a great interest, from Nordic and international scholars and collectors, in collecting Sámi remains, especially crania, in the nineteenth and early twentieth centuries, which led to an international exchange and trade with Sámi remains. Today, the extent of the exchange and trade with Sámi human remains and the present location of the remains are unclear. There also exist many contemporary accounts about local protests and resistance against the digging of Sámi graves, and of the highly unethical ways in which many excavations were conducted, often in secrecy (Ojala, 2009, 2016).

Homecomings

As a result of Sámi struggles for cultural rights, self-determination, and decolonization of the policies and structures of heritage management, several homecomings have taken place in Sápmi in recent times. Sámi activists, cultural workers and politicians have challenged museum and heritage institutions, and demanded the return of cultural objects and ancestral human remains to descendent communities, as part of cultural revitalization, repatriation and reburial processes.

In recent years, several reburial cases have taken place in Norway, Sweden and Finland, following demands from Sámi activists and local populations. Reburial cases are complex processes, and often challenging to carry out in practice, involving different actors and diverse ways of relating to the remains and the past.

One example of a Sámi reburial process is the reburial in Lycksele, northern Sweden, in 2019, when 25 crania were reburied. The ancestral remains were excavated in the 1950s from the old cemetery of Lycksele that had been used in the seventeenth and eighteenth centuries, and that had been established as part of the colonial missionary process. Lycksele was one of the church and market places that the Swedish Crown and Church established in the early seventeenth century, as mentioned above, where the local Sámi communities had to be present during certain periods of the year, in order to trade at the markets, pay taxes, attend court proceedings, go to church—and bury their deceased relatives at the churchyard.

An interesting process preceded the reburial, in which different actors—from Sámi organizations, the local community, museums, the Church and other institutions—collaborated to examine the history of the excavations, the attitudes of the researchers involved, how the human remains had been treated in the museums, and what effects the excavations had on the local population—within the framework of a reconciliation process (Aurelius, 2019). This process illustrates how repatriation entails so much more than only the transfer of objects or remains.

There was a large reburial ceremony at Lycksele, which also attracted a lot of media attention. Representatives from universities and museums carried the ancestral remains back and handed them over to Sámi community members who performed the reburial (the author took part in the ceremony as one of those representatives carrying the remains back to the grave). Representatives of the Swedish History Museum in Stockholm and the Swedish Church made public apologies for how the human remains had been treated. This ceremony was a very important event, and had great impact—in many ways, it changed how repatriation was understood by heritage authorities and institutions in Sweden.

A recent reburial case concerns Akamella cemetery, by the Muonio River (the border river with Finland) in northernmost Sweden. In 1878, a grave plundering took place at the cemetery, conducted by actors from Finland. The human remains were taken to the anatomical collections at the University of Helsinki (Ojala, 2022b). In the summer of 2024, a reburial of 23 crania was carried out at the cemetery. Akamella was used for burials by both the local Sámi population and the local Tornedalian population (the Meänkieli-speaking national minority population). The reburial ceremony was framed as an effort to unite the different groups, in a region with many local conflicts over land rights (the author also participated in the ceremony, as one of the people carrying the human remains to the grave).

Another example is the reburial in 2011 in Neiden, northern Norway, where remains from 94 individuals that had been excavated in 1915 from the local Orthodox Skolt Sámi cemetery, despite protests from the local community and congregation, were reburied. However, before the reburial there was a discussion in the Sámi community in Norway, with different views within the local Skolt Sámi community on whether the remains should be reburied at once or if there should be more research done first, in order to explore the local Skolt Sámi history, which is little known (Svestad, 2013). This example shows that reburial processes are complex, with actors with different wishes, interests and priorities in local communities, which should be heard in the debates.

A large reburial took place in northern Finland in August 2022, when human remains from about 140 individuals from the old anatomical collections at the University of Helsinki, which had been collected at the end of the nineteenth century and in the early twentieth century, were reburied in Inari/Aanaar, Utsjoki/Ohcejohka and Nellim/Njeä'llem.

Repatriation processes concerning Sámi ethnographical objects have also taken place in the Nordic countries, such as the Bååstede project in Norway, in which half of the Sámi ethnographical collections from Norsk Folkemuseum in Oslo were transferred to different Sámi museums in Norway (Gaup et al.,

2021), and the transfer of the Sámi collections from the National Museum in Helsinki, Finland, to Sámi Museum Siida in Inari/Aanaar (Nylander, 2023).

Revitalization processes in Sámi communities also include the use of old museum collections and the remaking of old objects, such as the Ládjogahpir. The use of the Ládjogahpir, a traditional Sámi women's hat, was suppressed as part of the colonial, missionary and assimilation processes. However, as part of revitalization and rematriation counter-processes, the hat has been produced and used by Sámi women once again, based on studies of old museum collections (Harlin & Pieski, 2020). Rematriation is used as a concept to describe how the returned heritage objects become active parts of the descendent communities—in the case of the Ládjogahpir also as craftivism (crafts + activism) and decolonial feminism (Nylander, 2023).

Archaeology and Sustainable Development

There is a great potential in archaeological approaches to sustainable development, not only to help understand colonial processes and colonial landscapes, but also to explore Indigenous landscapes, livelihoods and land use, and Indigenous self-governance before and during the colonial expansion.

The concept of biocultural heritage has been increasingly used in archaeological studies, aiming to overcome the rigid distinction between "nature" and "culture", as well as between human and non-human, and to explore the intertwinement of cultural heritage, landscape and place (Lane & Benny, 2024; cf. also Ferrara, 2024). Furthermore, recent years have seen growing discussions on historical ecology in archaeology, aiming to understand long-term human–environment relations and interactions.

Archaeology is a cross-disciplinary subject, encompassing aspects of the humanities, social sciences and natural sciences. Recent discussions on archaeology and climate change have focused on how archaeology can be part of discussions on how to handle climate change in the future, informed by deep-time historical perspectives (see, e.g., Britton & Hillerdal, 2019; Burke et al., 2021). Calls have been made to enhance the contribution of archaeology and heritage studies to the climate challenges faced by humanity (ICOMOS Climate Change and Cultural Heritage Working Group, 2019; Rockman & Hritz, 2020).

Recent historical and archaeological research has addressed Sámi self-governance and strategies for land use and management of natural resources, with a high degree of autonomy in the early modern period (see Norstedt, 2018; Larsson & Päiviö Sjaunja, 2022). These perspectives can be important for discussions on alternative forms of resource management, including management of common pool resources, in the northern areas—and provide evidence of Sámi Indigenous strategies to manage biocultural diversity.

Material and landscape sources can play an important role in these historical explorations. There is a potential of archaeology and material culture studies to throw light on aspects that are not covered by the written sources and to provide new perspectives. Archaeology can also serve as a method or an approach to explore local history, through material remains, places and landscapes. In community-based archaeological projects, local and Indigenous communities can be empowered to explore their own histories.

Internationally, the field of Indigenous archaeology has grown in importance in recent decades. Indigenous archaeology is a multifaceted movement, promoting Indigenous perspectives, experiences and knowledge in archaeology and heritage management, and working against colonial legacies and structures in science, museums and heritage institutions (Atalay, 2006; Ojala, 2019, 2022a; Nicholas & Watkins, 2025). Within Indigenous archaeology, archaeologists and community members have focused on developing community-based archaeology and heritage management, and collaborative, participatory approaches (Atalay, 2012). Issues of repatriation and reburial—as ways to confront and deal with colonial histories, and work for Indigenous self-determination—have also been central in the Indigenous archaeology movements in many parts of the world (Fforde et al., 2020; Meloche et al., 2021). Indigenous archaeology can be seen as socially engaged and transformative archaeology, with archaeologists working for changes in society (Atalay et al., 2014).

George Nicholas, an archaeologist from Canada who has worked extensively with Indigenous and community-based archaeology, has presented his view on the intersection of heritage and indigeneity in three basic points, which form a basis for Indigenous archaeologies:

1. That access to and control over one's own heritage is a basic human right essential to identity, well-being and worldview;
2. that Indigenous peoples in so-called 'settler countries' have historically been separated from their heritage, experienced little benefit from heritage-related research and suffered cultural harms and economic loss as a result; and
3. that community-based heritage initiatives are capable of challenging colonial structures in the research process without compromising the integrity of archaeology. (Nicholas, 2017, p. 200)

Approaches to Indigenous archaeology and community-based archaeology have not been so well developed in Swedish archaeology, especially not in contract archaeology (Knutson, 2024). However, there are several examples of collaborative community-archaeology approaches in Sápmi, especially in the South Sámi regions. Several collaborative, community-based survey and documentation projects, involving local communities, such as local Sámi villages, and professional archaeologists and museums, have been conducted (Ljungdahl & Norberg, 2012; Norberg & Winka, 2014; cf. also Knutson, 2021; Kroik, 2022). The South Sámi area is a region where Sámi archaeology has been particularly controversial, with several court cases dealing with the rights of local Sámi communities to reindeer grazing lands, in which archaeologists have also taken part as expert witnesses (Ojala, 2009; Brännström, 2020).

As Sápmi stretches across the present-day national borders, there are differences in heritage management systems between the different countries, and in the extent of Sámi self-determination in heritage issues.

There are Sámi Parliaments in Norway, Sweden and Finland, elected by registered Sámi constituencies, with varying degrees of influence and power in heritage management. In Norway, the Sámi Parliament has the most power. It is responsible for Sámi heritage management, and also has control over collections of Sámi ancestral remains, and employs own archaeologists and cultural specialists. In Finland, the Sámi are claiming a stronger position in heritage management. Also, there have been very heated discussions in Finland on the definition of Sáminess, who is a Sámi, and who should be allowed to vote to the Sámi Parliament (Lehtola, 2015; Valkonen, 2019). At present, a new heritage legislation in Finland is being developed, with a vision of developing a clearer implementation of Sámi cultural autonomy in Finland. In Sweden, the Sámi Parliament has little power. There is only one, uniform state system of heritage management in Sweden. However, discussions on future Sámi heritage management have begun (Knutson, 2024). In Russia, there is no Sámi Parliament, and the current situation is very difficult for Indigenous groups, with political repression and difficulties in collaborating with Indigenous groups in other parts of the world (cf. Danilov, this volume).

Dealing With Colonialism?

In the Nordic countries, the general recognition of and public debates on Nordic colonialism in general and colonial history in Sápmi in particular have increased in recent years.

Truth commissions have been initiated by the governments in Norway, Sweden and Finland, dealing with the historical relations between the states and the Sámi people, for instance assimilation, discrimination, school and language policies, and racial research. In Norway, the Truth commission has already published its final report (Sannhet og forsoning, 2023), while the commissions are ongoing in Sweden and Finland.

The Church of Sweden has started a reconciliation process, and recently made a public apology to the Sámi people (Lindmark & Sundström, 2016). The Church of Sweden has also been very engaged in many of the reburial cases mentioned above. The Swedish government has, furthermore, established a truth and reconciliation commission concerning the Tornedalian, Kven and Lantalaiset populations in the county of Norrbotten in northernmost Sweden. The Tornedalian population is an officially recognized national minority population in Sweden.

There are, of course, many challenges connected with these commissions—especially when the goal is to facilitate reconciliation. It remains to be seen whether the commissions will lead to concrete results in actually changing state policies and practices. This will depend very much on the current political situation in the countries (Kuokkanen, 2020; Szpak & Bunikowski, 2021).

In debates on Sámi land and cultural rights, International law and discourses on human rights play an important role. The United Nations Declaration on the Rights of Indigenous Peoples (UNDRIP), from 2007, is especially relevant in this context. Several articles in the Declaration deal with Indigenous cultural rights. For instance, Article 12.1 states that:

> Indigenous peoples have the right to manifest, practice, develop and teach their spiritual and religious traditions, customs and ceremonies; the right to maintain, protect, and have access in privacy to their religious and cultural sites; the right to the use and control of their ceremonial objects; and the right to the repatriation of their human remains. (UNDRIP, 2007)

However, for a long time it has been quite unclear how these rules should be applied in practice within heritage management in Sweden. Recently, the Swedish National Heritage Board has issued new guidelines for the handling of human remains in museum collections and the handling of repatriation claims, which recognize—for the first time in Sweden—the relevance of UNDRIP in dealing with Sámi heritage (RAÄ, 2020a, 2020b). A larger discussion concerning Sámi self-determination and the recognition of Sámi rights in heritage management in Sweden is still needed.

As has been discussed above, quite a lot is happening right now in Sweden and the Nordic countries concerning the repatriation and reburial debates. An earlier reluctant or negative attitude of many museums and heritage institutions in Sweden has gradually changed towards a more open, understanding and positive attitude to Sámi perspectives on repatriation and reburial. Moreover, in recent years, Sámi culture has attracted much attention in Sweden—Sámi art, handicraft (*duodji*), literature, music and films have become very popular and highly esteemed. But what do the recent developments actually mean for Sámi Indigenous rights, and Sámi self-determination?

At the same time as Sámi repatriation and reburial demands have become more and more accepted by museums, heritage institutions and government agencies, there is also another parallel development. There is a continuing, ever increasing pressure on Sámi lands and land use, from large-scale mining and plans for new mines and industrial developments, widespread and destructive forestry, the development of huge windfarms, and different infrastructural projects—today to a large extent framed as part of the so-called green transition, aiming to achieve a sustainable future. These pressures are adding to the earlier extensive damming of rivers and lakes for water power, which has flooded huge areas of land. There are growing conflicts over land rights in Sápmi, reindeer grazing lands and migration routes, as well as hunting and fishing rights. The Sámi voices and the Sámi struggle to protect land rights and traditional land use are most often not met with the same understanding as Sámi demands for cultural rights in recent years. So how should we understand the current situation in Sápmi with these parallel, different developments?

Conclusion

In this chapter, I have argued that there is a need for historical perspectives in debates on sustainable development and Indigenous and minority rights in Sápmi, and elsewhere—and that archaeology and heritage studies can play a role in these debates. Archaeological engagements with places and landscapes, and the traces of earlier times in the ground, have the potential to give voice to people, places, things and events which were never recorded in the historical records.

The development of an "extractive mindset", as exemplified by the discussion above on the exploitation of natural resources, early modern collecting of Sámi material culture, and nineteenth- and twentieth-century collecting of Sámi human remains, constitutes an important part of the historical contextualization of present-day debates on extractive industries, Indigenous rights and sustainable development in Sápmi.

The repatriation, rematriation and reburial processes discussed in this chapter illustrate how in recent times the direction of the flow of heritage has been turned around and how Sámi perspectives have been included in discourses of heritage management. This represents an alternative manner of dealing with colonial heritage, which is more attentive to local and Indigenous experiences and perspectives, and which expresses an alternative vision of sustainable heritage management.

Still, there is currently a great need to critically discuss how to deal with Sámi colonial history and heritage in the future, including museum collections in the Nordic countries and elsewhere. There is today a strong colonial legacy in Sápmi, as demonstrated by the ongoing exploitation of natural resources—such as mining and wind power development—and the Sámi struggles for protection of lands and traditional livelihoods.

In archaeology and heritage management, there has been a lack of perspectives from Sámi and local communities, which has led to imbalances and inequality. The power dynamics in archaeology and heritage management need to be critically discussed, as well as the responsibilities of scholars, and the roles of archaeology, heritage management and museums.

There is great potential in developing community-based and community-initiated heritage research and management, as part of a larger work to develop alternative approaches to the implementation of the Sustainable Development Goals, turning away from the long-term extractive mindset. Archaeological approaches can contribute to debates on sustainable development, and to the inclusion of multiple perspectives on sustainability. Exploring the deep historical roots of present-day issues and conflicts relating to land and cultural rights is one step in decolonization efforts, and can provide perspectives on the complexities of interrelations between past and present, and possible future sustainable relations in our world.

References

Atalay, S. (2006). Indigenous archaeology as decolonizing practice. *American Indian Quarterly, 30*(3–4), 280–310.

Atalay, S. (2012). *Community-based archaeology: Research with, by, and for Indigenous and local communities.* University of California Press.

Atalay, S., Clauss, L. R., McGuire, R. H., & Welch, J. R. (2014). *Transforming archaeology: Activist practices and prospects.* Left Coast Press.

Aurelius, A. (2019). *Máhtsatiebmie likttemijne—Återbördande i försoning: En processbeskrivning över repatrieringen i Lycksele 2019.* Liksjuon Sámiensiäbrrie.

Bergling, R. (1964). *Kyrkstaden i övre Norrland. Kyrkliga, merkantila och judiciella funktioner under 1600- och 1700-talen.* Almqvist & Wiksells.

Brännström, M. (2020). The enhanced role of archaeological and historical research in court proceedings about Saami land rights. In M. Spangen, A.-K. Salmi, T. Äikäs, & M. Fjellström (Eds.), *Currents of Saami pasts: Recent advances in Saami archaeology* (pp. 177–188). Archaeological Society of Finland.

Britton, K., & Hillerdal, C. (2019). Archaeologies of climate change: Perceptions and prospects. *Études/Inuit/Studies, 43*(1/2), 265–288.

Burke, A., Peros, M. C., Wren, C. D., Pausata, F. S. R., Riel-Salvatore, J., Moine, O., de Vernal, A., Kageyama, M., & Boisard, S. (2021). The archaeology of climate change: The case for cultural diversity. *PNAS, 118*(30), 1–10.

Ferrara, V. (2024). *Historical olive agroecosystems of Sicily: Operationalising biocultural heritage for sustainable futures.* Uppsala University.

Fforde, C., McKeown, C. T., & Keeler, H. (2020). *The Routledge companion to Indigenous repatriation: Return, reconcile, Renew.* Routledge.

Fredriksson, A., Nordin, J. M., & Ojala, C.-G. (2019). The Deductionis coloniarum rationes & causae (1668): Building arguments for Swedish colonialisation. *Lias, 48*(1), 123–193.

Gaup, K. E., Jensen, I., & Pareli, L. (2021). *Bååstede: The return of Sámi cultural heritage.* Museumsforlaget.

Harlin, E.-K., & Pieski, O. (2020). *Ládjogahpir: Máttaráhkuid gábagahpir.* Davvi Girji.

ICOMOS Climate Change and Cultural Heritage Working Group. (2019). *The future of our pasts: Engaging cultural heritage in Climate action.* ICOMOS.

Knutson, C. (2021). *Conducting archaeology in Swedish Sápmi: Policies, implementations and challenges in a postcolonial context.* Linnaeus University.

Knutson, C. (2024). *Indigenous archaeology in Sweden: Aligning contract archaeology with national and international policies on Indigenous heritage.* Linnaeus University.

Kroik, Å. V. (2022). *Dihte gievrie—det vi möter i respekt: Berättelser om en sydsamisk trumma.* Uppsala University.

Kuokkanen, R. (2020). Reconciliation as a threat or structural change? The truth and reconciliation process and settler colonial policy making in Finland. *Human Rights Review, 21*(3), 293–312.

Lane, P. J., & Benny, Q. S. (2024). Biocultural heritage: Definitions, applications, and African case studies. In A. Sinamai, J. D. Giblin, S. Chirikure, & I. Odiaua (Eds.), *Routledge handbook of critical African heritage studies* (pp. 323–346). Routledge.

Larsson, J., & Päiviö Sjaunja, E.-L. (2022). *Self-governance and Sami communities: Transitions in early modern natural resource management.* Palgrave Macmillan.

Lehtola, V.-P. (2015). Sámi histories, colonialism, and Finland. *Arctic Anthropology, 52*(2), 22–36.

Lindmark, D., & Sundström, O. (2016). *De historiska relationerna mellan Svenska kyrkan och samerna* (Vol. I and II). Artos & Norma.

Ljungdahl, E., & Norberg, E. (2012). *Ett steg till på vägen: Resultat och reflexioner kring ett dokumentationsprojekt på sydsamiskt område under åren 2008–2011.* Gaaltije – sydsamiskt kulturcentrum.

Meloche, C. H., Spake, L., & Nichols, K. L. (2021). *Working with and for ancestors: Collaboration in the care and study of ancestral remains.* Routledge.

Nicholas, G. (2017). Culture, rights, indigeneity and intervention: Addressing inequality in Indigenous heritage protection. In A. Karlström, C. Hillderdal, & C.-G. Ojala (Eds.), *Archaeologies of 'us' and 'them': Debating history, heritage and indigeneity* (pp. 199–217). Routledge.

Nicholas, G., & Watkins, J. (2025). *Working as Indigenous archaeologists: Reckoning new paths between past and present lives.* Routledge.

Norberg, E., & Winka, U. S. (2014). *Sydsamer—landskap och historia: ett dokumentationsprojekt på sydsamiskt område under åren 2012–2014.* Gaaltije – sydsamiskt kulturcentrum.

Nordin, J. M., & Ojala, C.-G. (2019). Collecting, connecting, constructing: Early modern commodification and globalization of Sámi material culture. *Journal of Material Culture, 23*(1), 58–82.

Norstedt, G. (2018). *A land of one's own: Sami resource use in Sweden's boreal landscape under autonomous governance.* Swedish University of Agricultural Sciences.

Nylander, E.-K. (2023). *From repatriation to Rematriation: Dismantling the attitudes and potentials behind the repatriation of Sámi heritage.* University of Oulu.

Ojala, C.-G. (2009). *Sámi prehistories: The politics of archaeology and identity in northernmost Europe.* Uppsala University.

Ojala, C.-G. (2016). Svenska kyrkan och samiska mänskliga kvarlevor. In D. Lindmark & O. Sundström (Eds.), *De historiska relationerna mellan Svenska kyrkan och samerna* (pp. 993–1028). Artos & Norma.

Ojala, C.-G. (2019). Colonialism past and present: Archaeological engagements and entanglements. In T. Äikäs & A.-K. Salmi (Eds.), *The sound of silence: Indigenous perspectives on the historical archaeology of colonialism* (pp. 182–201). Berghahn Books.

Ojala, C.-G. (2020). Mines and missions: Early modern Swedish colonialism in Sápmi and its legacies today. In M. Spangen, A.-K. Salmi, T. Äikäs, & M. Fjellström (Eds.), *Currents of Saami pasts: Recent advances in Saami archaeology* (pp. 160–176). Archaeological Society of Finland.

Ojala, C.-G. (2022a). Indigenous archaeology. In M. Lindroth, H. Sinevaara-Niskanen, & M. Tennberg (Eds.), *Critical studies of the Arctic: Unravelling the north* (pp. 99–122). Palgrave Macmillan.

Ojala, C.-G. (2022b). *Insamling av mänskliga kvarlevor från Tornedalen: Rapport till Sanningskommissionen för tornedalingar, kväner och lantalaiset (ku 2020:01).* Regeringskansliet.

Ojala, C.-G. (2023). Sámi Mobilities in colonial spaces and the right to make a home. *Journal of Imperial and Commonwealth History, 51*(3), 576–597.

Ojala, C.-G., & Nordin, J. M. (2015). Mining Sápmi: Colonial histories, Sámi archaeology, and the exploitation of natural resources in northern Sweden. *Arctic Anthropology, 52*(2), 6–21.

Ojala, C.-G., & Nordin, J. M. (2019). Mapping land and people in the north. Early modern colonial expansion, exploitation and knowledge. *Scandinavian Studies, 92*(1–2), 98–133.

RAÄ. (2020a). = God samlingsförvaltning: Stöd i hantering av mänskliga kvarlevor i museisamlingar. *Riksantikvarieämbetet.*
RAÄ. (2020b). = God samlingsförvaltning: Stöd för museer i återlämnandeärenden. *Riksantikvarieämbetet.*
Rockman, M., & Hritz, C. (2020). Expanding use of archaeology in climate change response by changing its social environment. *PNAS, 117*(15), 8295–8302.
Sannhet og forsoning. (2023). = *Sannhet og forsoning: Grunnlag for et oppgjør med fornorskingspolitikk og urett mot samer, kvener/norskfinner og skogfinner.* Rapport til Stortinget fra Sannhets- og forsoningskommisjonen.
Snickare, M. (2022). *Colonial objects in early modern Sweden and beyond: From the Kunstkammer to the current museum crisis.* Amsterdam University Press.
Svestad, A. (2013). What happened in Neiden? On the question of reburial ethics. *Norwegian Archaeological Review, 46*(2), 194–242.
Szpak, A., & Bunikowski, D. (2021). Saami truth and reconciliation commissions. *The International Journal of Human Rights, 26*(2), 306–331.
UNDRIP. (2007). = *the United Nations declaration on the rights of indigenous peoples.* http://www.un.org/esa/socdev/unpfii/documents/DRIPS_en.pdf
Valkonen, S. (2019). Conceptual governance on defining indigeneity: The Sámi debate in Finland. In T. H. Eriksen, S. Valkonen, & J. Valkonen (Eds.), *Knowing from the indigenous north. Sámi approaches to history, politics and belonging* (pp. 142–162). Routledge.

CHAPTER 3

Beyond Melting Ice: Decolonising Environmental Governance in the Arctic

Mark Nuttall

Introduction

Decolonising the Sustainable Development Goals (SDGs) calls for a critical re-evaluation and rethinking of them to ensure that they are inclusive of and responsive to the needs, perspectives, and knowledge systems of Indigenous and marginalised communities. This process requires recognising the historical and ongoing impacts of colonialism and other social, economic, and political processes on these communities and addressing the power imbalances that perpetuate their exclusion from decision-making. And if the SDGs are to be effective, their achievement will not be possible without a shift from a top-down, often largely technocratic approach to sustainable development and environmental management, to one that is participatory and co-created with those who are most affected by sustainable development policies. To do so involves meaningful engagement with and integration of Indigenous knowledge systems and practices—not just merely acknowledging their importance—into local, regional and global frameworks. Furthermore, decolonising the SDGs necessitates a focus on what justice and reparative action would look like. This is not an easy task in practice—not least because of the methodological difficulties and uncertainties involved in how to go about actually doing and enacting collaborative resource management, but also because it makes demands of those responsible for implementing the SDGs to acknowledge and address the disproportionate burdens of environmental degradation and climate change borne by Indigenous and marginalised communities.

M. Nuttall (✉)
Department of Anthropology, University of Alberta, Edmonton, Canada
e-mail: mark.nuttall@ualberta.ca

R. Toivanen et al. (eds.), *Decolonizing the Sustainable Development Goals*,
https://doi.org/10.1007/978-3-032-14263-4_3

SDGs 13 and 14—Climate Action and Life Below Water—have significance for Inuit communities in northern Greenland and Arctic Canada, High Arctic areas that have been the focus of much of my long-term research as a social anthropologist. In this chapter, I consider the place of ice and Arctic ecosystems in conservation discourse, which draws heavily on images and vocabulary that emphasise ecological precarity. For example, images of collapsing glaciers and emaciated polar bears circulating in global media evoke a new form of terror—an anxiety about impending planetary catastrophe. Such representations of the Arctic are difficult barriers to break down. They often privilege the non-human and imaginaries of the spectacular, and they advance particular environmentalist and conservationist agendas while sidelining Indigenous perspectives. This selective framing, I argue, risks reinforcing narratives of Arctic exceptionalism while oversimplifying the entanglements of complex socio-ecological relationships.

However, I point to the ways Inuit respond to this selective framing through Indigenous initiatives for dealing with the effects of climate change, for the introduction of environmental protection initiatives, and for regulating shipping, commercial fishing, and industry. These are most obvious in the form of initiatives for co-management, community-based monitoring, and Indigenous Protected Areas (IPAs) that recognise Indigenous knowledge as fundamental to conservation and environmental management and foreground Indigenous rights. Indigenous people's organisations, such as the Inuit Circumpolar Council (ICC), work to ensure environmental governance and conservationist narratives take serious note of Indigenous needs, including rights to hunt and fish, and perspectives on water and ice, as well as Indigenous knowledge of human and non-human entanglements.

Arctic Melt

Over the last few years, I have worked with communities in northern Greenland and with colleagues from the natural and physical sciences, such as oceanographers, glaciologists, and marine geologists, to understand the changing nature of sea ice, glaciers, and the marine environment under conditions of rapid climate change (Nuttall, 2017, 2024; Ribeiro et al., 2021). I have also been concerned with understanding how northern Greenland, Canada's Arctic Archipelago, and the Arctic Ocean have become defined by scientists, conservationists, and environmental organisations as spaces where urgent action is needed to meet the challenges posed by climate change. Scientists warn that the Arctic Ocean's ice cover is approaching a tipping point and may disappear completely during summer months over the next two to three decades (e.g., Foss et al., 2024; Jahn et al., 2024; Notz & Stroeve, 2018). They also document rapid and dramatic ice-mass loss of Greenland's inland ice and the retreat of tidewater glaciers, which is happening with greater speed in certain coastal areas (e.g., Millan et al., 2023). Conservation and environmental groups commonly frame the Arctic as a critical zone of planetary crisis. Organisations such as the World Wide Fund for Nature (WWF) and Oceans North, for instance, have launched campaigns to protect threatened northern waters, last areas of ice, polynyas, polar bears, and narwhals, with the Last Ice Area now entering mainstream scientific discourse and scenario making (Adams, 2014; Newton et al., 2021).

Thinning sea ice, melting glaciers and ice sheets, and threatened habitat for Arctic species enter the global imagination as the effects of climate change, provoking anxieties over the disappearance of Arctic ecosystems and wildlife. Similarly, environmentalist action and conservation policy tend to hinge on ideas of ruination, loss, and absence in a future Arctic. Polar bears and other wildlife that conservationists often think of as charismatic, such as whales and narwhals, become central to the most influential discourses about biodiversity and environmental change. These shape regional and global approaches to marine conservation and resource development in the region and, I argue, reinforce and simultaneously continue to inform ideas about the sustainable use of marine resources and mainstream notions of sustainable development.

It would be an unfair criticism to say that scientists, conservationists, and environmentalists are not well intentioned in how they raise awareness of a melting Arctic and advocate for the protection of disappearing ice, the establishment of wildlife preserves, and the introduction of stricter conservation measures for animal and fish species that form the basis of Indigenous hunting and fishing livelihoods. For many Indigenous residents who depend on the living resources of sea and land, life seems increasingly uncertain, precarious, and risky. They are also caught up in an economic and geopolitical scripting about resource frontiers of potential threats to global security. However, some of these interventions and conservation strategies continue to apply logics that tend to be informed and legitimised by scientific approaches, observation and monitoring, and valuations of wildlife and ecosystems that more often than not work, even if unwittingly, to exclude the perspectives of Inuit communities. Although WWF and Oceans North do engage with—and are often supportive of—Inuit communities (Oceans North, a Canadian organisation, has opened an office in Greenland, staffed by Greenlanders, and collaborates with the ICC), these are not always friction-free partnerships to forge and sustain. Arctic ecosystems and biodiversity, as is true for other ecosystems, remain subject to a technocentric approach to conservation and environmental management strategies (Villas Boas & Marçal, 2022). Furthermore, conservation organisations often proceed from a perspective that privileges the environment and animals, identifying specific ecosystems as being at risk, and elevating some species to the status of emblems of a changing Arctic, as I will show. People living in High Arctic coastal communities, however, are challenging such action, especially in how they respond to claims made about who is entitled to speak for the non-human.

SDG 13, which focuses on climate action, can be critiqued for its often top-down approach that can marginalise Indigenous and local communities. One key criticism is that the strategies and policies devised to combat climate change are frequently developed without adequate consultation with the people most affected by these policies. This can lead to interventions that are misaligned with local realities and needs. For instance, global climate initiatives tend to emphasise carbon trading, renewable energy technologies, and large-scale conservation projects, which may overlook or undermine traditional practices and local forms of environmental management. As Astrid Ulloa (2017, 2019) argues, global climate policies are predominantly informed by Western scientific paradigms, which often exclude Indigenous perspectives and knowledge systems. This epistemic dominance perpetuates a form of environmental governance that prioritises technocratic solutions over culturally grounded, community-driven approaches. To address these imbalances, it is crucial to incorporate Indigenous worldviews into policy frameworks, ensuring that local communities have meaningful participation. By not fully engaging with these knowledge systems, the implementation of policies that aim to reach Goal 13 could end up inadvertently reinforcing existing disparities and fail to address the root causes of climate vulnerability in marginalised communities.

Similarly, SDG 14, which focuses on the conservation and sustainable use of oceans, seas, and marine resources, has faced critiques for its tendency to prioritise biodiversity conservation over the livelihoods and rights of local fishing communities. Marine conservation initiatives can often lead to the establishment of marine protected areas (MPAs) that restrict access to traditional fishing grounds, thereby threatening the food security and economic stability of coastal communities (Bennett et al., 2015). Overall, critiques of Goals 13 and 14 emphasise the need for a more inclusive and equitable approach to climate action and marine conservation. This involves recognising and integrating Indigenous and local knowledge systems, ensuring meaningful participation of affected communities in decision-making processes, and addressing the socio-economic impacts of environmental policies. Büscher and Fletcher (2020), for example, propose a model of "convivial conservation", which advocates for breaking down the dichotomy between humans and nature. Convivial conservation, as they articulate it, seeks to integrate conservation into the everyday lives of local communities, rather than treating it as a separate domain requiring exclusionary practices. This approach aligns closely with decolonial ideals, as it emphasises respect for local ways of life, knowledge, experience, and governance approaches (see also, Buschman & Sudlovenick 2023).

Spectacular Ice

In the nineteenth century, during a time of increased Euro-American activities of whaling, mapping and discovery, the Arctic was often depicted in accounts of exploration and in literature as an inhospitable, frozen region where glaciers, icebergs, and sea ice were seen as daunting obstacles to navigation and movement. The grandeur and majesty of the ice-covered northern world were considered both beautiful and deadly. This perspective was prevalent among non-Indigenous visitors, such as explorers, whalers, adventurers, and scientists, who travelled north primarily during the spring and summer months. They wrote about the ice they encountered as omnipresent, unchanging, and forbidding. Terror and horror were common themes in the narratives of northern voyages and polar expeditions, which often emphasised the harrowing experiences of being trapped in the ice for months on end. Accounts of voyages in search of the Northwest Passage, for instance, often highlighted the overwhelming and deadly challenges posed by sea ice (Williams, 2009), while artists such as Caspar David Friedrich and William Bradford created works that embodied the sublime and terrifying nature of the Arctic, illustrating human vulnerability against the formidable forces of nature (Hill, 2009).

Further elaboration of this theme can be seen in the work of Potter (2007), who highlights how European imaginaries constructed the Arctic as a space of desolation and danger, contributing to a colonial gaze that rendered the region as both a site of adventure and a zone to be conquered. Similarly, McCorristine (2018) explores the spectral and haunting qualities attributed to the Arctic, connecting these with anxieties over ecological loss in contemporary times. These literary and artistic depictions played a role in shaping public and scientific perceptions of the Arctic during that period and continue to influence conservationist narratives today.

The nineteenth-century mediascape (cf. Appadurai, 1990) was rich with books about exploration, evocative paintings, poignant poetry, and maps that featured vast, blank spaces, symbolising the unknown and uncharted territories of the Arctic. These cultural artefacts allowed European and North American audiences to experience the mysteries and dangers of the Arctic vicariously. While explorers' accounts and artistic and literary depictions of the Arctic were primarily concerned with the challenges of human exploration and survival, in contemporary times feelings of terror and horror associated with the Arctic remain. However, they have shifted, influenced by modern scientific research and the pressing realities of climate change. The fear of physical entrapment in the ice may have diminished as non-Indigenous awareness and experience of the region and its increasing accessibility means more people visit each year, including on 'expedition'-styled cruise ships, yet the Arctic now symbolises a different kind of dread. The scientific documentation of climate change effects, such as the melting of sea ice and glaciers and the ecological crises that follow, has profoundly impacted public perceptions. Images of polar bears stranded and starving on shrinking ice floes and dramatic footage of glaciers collapsing and receding have replaced the nineteenth-century (and later) depictions of the Arctic sublime. These contemporary images evoke a deep sense of horror and foreboding, not because of the Arctic's physical challenges, but due to the realisation of an impending global ecological catastrophe. Their deployment is a crucial technique for influencing and shaping public and policy maker perceptions of climate change and species extinction in the Arctic. These images are powerful tools in communicating the urgency of environmental crises, and they are intended to invoke feelings of awe, horror, and moral responsibility.

Yet by centring spectacular images, the risk is that a single narrative is created that simplifies the multiplicity of Arctic realities.

Decolonising Conservation

Decolonising conservation involves rethinking and restructuring conservation practices to prioritise the rights, knowledge, and needs of Indigenous and local communities that have historically been marginalised by traditional conservation efforts. Traditional conservation is characterised by policies and strategies such as those that establish protected areas and reserves that exclude human activity, frequently leading to the displacement of Indigenous peoples and restricting the activities of local communities (Dawson et al., 2023a, b). For example, in Papua New Guinea, Paige West (2006) demonstrated how international conservation NGOs often implement policies that clash with local cultural practices, leading to tensions and failures in achieving conservation objectives. This approach is critiqued for failing to recognise the sustainable land management practices of Indigenous communities (Adams & Mulligan, 2003). West (2006) argues that a decolonial approach must prioritise the sovereignty and self-determination of Indigenous peoples, ensuring that conservation is not a top-down process imposed by external actors but a collaborative effort driven by local communities. To decolonise conservation is to advocate for inclusive models such as collaborative resource management and community-based conservation, which integrate Indigenous knowledge systems and governance structures. These models emphasise the need for conservation policies that respect Indigenous land rights, cultural practices, and autonomy, fostering more sustainable and just environmental stewardship (Büscher and Fletcher ibid.).

Scholars like Asiyanbi et al. (2019) and Davis and Todd (2017) argue that conservation projects, environmental initiatives, and conceptualisations of our modern age as the Anthropocene often reflect and perpetuate colonial power dynamics, where decisions are made by international organisations and governments without adequate consultation or involvement of Indigenous people and local communities. Decolonising conservation, then, is a call for the creation of spaces for Indigenous voices and leadership in conservation discourse and practice. This includes not only acknowledging Indigenous contributions to biodiversity but also supporting their sovereignty and self-determination. By shifting towards models that value and incorporate Indigenous perspectives, conservation efforts can become more effective and equitable, ensuring that those who are most connected to and dependent on land, water, and animals are also their stewards and decision-makers for effective protection and use.

Jim Igoe (2017) argues that spectacles (and spectacular ideas) of nature act to variously connect or disconnect the experiences people have of their surroundings or places that are distant to them. Images of threatened environments and endangered species are key to how people's perceptions of the world are shaped. They also have implications for the ways conservation can influence and disrupt human–environment relations. Similarly, Ursula Heise (2016) illustrates how the cultural production of endangered species has profound implications for the separation of the human and non-human in conservation discourse and practice. Walker DePuy et al. (2022) call for reflection on and critical examination of the concepts and practices that are rendered visible and invisible in current environmental governance paradigms and pursuits. They point out that many scholars, including Indigenous practitioners, have emphasised that the ideologies and enactments of coloniality are not only very much alive today in both knowledge and structural power relations, but live on through the dominant discourses, practices, and epistemologies of environmental governance. This echoes work by, for example, Mario Blaser (2016), whose research on how Innu in Labrador talk about caring for caribou points to ways of thinking about how a multiplicity of worlds that are animated in different ways troubles and challenges conservation ideologies and practices and, in the process, the relationships between Indigenous peoples and scientists—and how environmental governance can fall short if it ignores Indigenous ontologies concerning being, acting, and living in relation to the non-human.

Just as Michael Schnegg (2021) discusses in the case of climate change research in the arid regions of northwestern Namibia, such a scientific encounter with the world is enframed and mediated by technology, whereas Indigenous experiences are enmeshed with daily activities in surroundings that are comprised of human and non-human entities. At a broader level, global action against climate change continues to be backed by a single type of knowledge, illustrated by the ways policies are informed by IPCC reports that interpret climate through scientific constructions, understandings, and interpretations. As Ulloa argues, "...this global knowledge not only has political implications but has a local impact because it is implemented via a global policy that does not include local worldviews and knowledges" (Ulloa, 2019, p. 68). At the core of such critique is the notion that global climate governance frameworks often rely on a universalist approach that excludes local epistemologies. This exclusion results in policies that may have scientific authority behind them but are culturally and socially detached from the realities of affected communities. To bridge this gap, it is necessary to move beyond the homogenisation of knowledge and create spaces where multiple worldviews can co-exist and inform policy.

This resonates with my work in the Arctic. Melting sea ice may allow the smoother flow of shipping through northern waters and enable greater access to what have been regarded as previously remote Arctic regions, but it impedes the mobility of Indigenous peoples such as those Inuit hunters and fishers who rely on sea ice and glacial ice for travel routes and on the ice-dependent animal species they catch. For scientists with whom I work, and who monitor Arctic climate change, this reshaping of the world's high latitudes is often revealed as a view looking down from above the Earth through satellite images. However, on the ground, on the ice and on the water, people observe and experience these transformations in ways that are immediate, sensorial, and embodied.

And, significantly, they may not always think of the melting ice as the most pressing problem they face, nor do they necessarily worry about it as a sign of something apocalyptic. Transformative, yes, but also entirely consistent with Indigenous understandings and experiences of the world as always being in a process of becoming, and which guide ways of being in rather than adapting to it (Nuttall, 2019, 2024).

The Last Ice

The narrative of exceptionalism that informs the geopolitics of and about the Arctic is reinforced in the ways ice, animals, and ecosystems are represented as threatened in global imaginaries. This is illustrated by Greenpeace's Save the Arctic campaign and by the WWF campaign to designate coastal stretches of northern Greenland and the Canadian Eastern Arctic and parts of the Arctic Ocean as the Last Ice Area. WWF argues that this area will be the only part of the Arctic marine environment to retain its summer sea ice by 2050 (see: https://www.arcticwwf.org/our-priorities/last-ice-area/). The Last Ice Area initiative draws—quite rightly—global attention to the loss of sea ice in a threatened ecoregion. It is still a place, WWF argues, where sea ice and Arctic ecosystems are intact—but they are threatened by climate change and industrial activity—and so urgent international action is needed to protect these last remaining stretches. It resembles the ideas that sustain strategies for fortress conservation. It is in this way that the protection of ice has become key for ecological restoration in the Anthropocene Arctic. The Last Ice Area is described as a place that holds the promise for the formation and growth of new sea ice to replace that which has been lost. Calls for the protection of what remains are not underpinned by ideas of novel ecosystems and the emergence of new species assemblages, however. WWF posits that protecting this last area of ice is the only chance left to reverse the Arctic Ocean's trajectory towards a tipping point.

The image of the polar bear is key to the Last Ice Area campaign, which creates a very specific and particular wildlife-centric image of the Arctic in which people—most tellingly, Inuit hunters and their families who depend on catching and eating seals, whales, and polar bears—are relatively absent. Text, images, and films act together to construct, narrate, and disseminate a story that minimises and even ignores the presence and hunting rights of Indigenous people while emphasising the iconic nature of charismatic wildlife to viewers outside the Arctic. As a form of spectacular environmentalism, the invocation of 'the' Last Ice Area has a certain power in how it appeals to governments and a global public by emphasising, with great urgency, that the region is in crisis and is undergoing rapid melt. It also implies that the area is the only one of its kind and, in doing so, it reinforces ideas of Arctic exceptionalism and extends ideas of spectral presence and absence beyond species to ice as material substance.

Historian Shane McCorristine (2018) argues that such long-held associations continue to inform a public imagination in which the Arctic is enchanting, mysterious, magical, and even spectral. Its ruination and impending ecosystem collapse demands global intervention to save it. In an insightful article McCorristine wrote with geographer Bill Adams (McCorristone and Adams 2020), attention is drawn to the implications of a sense of the spectral for the way we think about conservation. The process of disappearing—and the state of disappearance—provides a powerful frame for conservation thinking and action. In the process, this allows for the threat of extinction, loss, death, and disappearance to generate lively, anxious, and provocative responses.

During my fieldwork over the last decade, I have listened to people in northwest Greenland talk about how they feel that conservationists, environmentalists, and the biologists who visit to study narwhals, polar bears, seals, and fish are more concerned with what is happening to wildlife and regional ecosystems than with the place of Inuit within the discussions taking place about biodiversity and conservation. The hunters with whom I work tell me that, of course, they agree that it is necessary to protect the ice on which polar bears, narwhals, and other animals, as well as human communities, depend. Yet they remain suspicious about the real intentions of organisations such as WWF, as well as those of biologists from Nuuk, Denmark, and elsewhere. They believe these intentions to be driven by a concern with protecting animals and fish and to have little regard for Inuit hunting and fishing rights. While WWF acknowledges the importance of ensuring sustainable communities in a changing Arctic, the worry for some people living in northwest Greenland is that the Last Ice Area initiative could privilege ice and polar bears over Indigenous rights to hunt.

It is marine biologists—including those who do monitoring work at the Greenland Institute of Natural Resources in Nuuk—who provide the data the Greenlandic authorities need to decide on quotas for the hunting of certain species of marine mammals and fish. And it is the hunting of narwhals that has become an especially contentious and heated issue over the last few years. The view presented by scientists is that narwhal hunting is not sustainable and that, combined with the effects of climate change, local hunting practices threaten northern Greenland's sub-populations of narwhal. They call for a form of targeted conservation, without local input, that would manage hunting—or even place a moratorium on it. Hunters, on the other hand, argue that scientists direct their focus on harvesting activities without attempting to understand the cultural and economic importance of hunting, the agency of narwhals, the ways they think and move around in the sea, and how people and narwhals in northwest Greenland interact and interrelate. In many instances, those scientists who are tasked with acquiring data on narwhals reject the consideration that local perspectives and worldviews could be useful (Heide-Jørgensen et al., 2020).

With other anthropologists, I have been caught up in this debate. When we have suggested the importance of bringing science and local knowledge into conversation, we have been accused of failing to understand the nature of Greenland's coastal ecosystems—and Greenland's resource governance system—and serving as apologists for hunters, in both north and east Greenland, whose real incentive to go after narwhals, it is claimed, is commercial and market driven (Heide-Jørgensen et al. ibid.). It is a familiar clash between Indigenous knowledge, science, and the bureaucracies of wildlife management, which highlights a disjuncture between the epistemological and the ontological, between ways of knowing, thinking, acting, and being.

Domains of conservation and environmental management offered as a solution to Arctic meltdown and environmental crisis can be biopolitical, in the way that Biermann and Anderson (2017) consider approaches to the management of endangered species and protected areas to be an entanglement of logics and techniques that assert power, but are overlapping and contradictory, and subject to multiple forms of biopolitics (Hodgetts, 2017). They are also geopolitical in how ideas of the environment, endangered species, and biodiversity are deployed to influence the ways state governments and publics imagine and think about nature under threat. The non-human, the things labelled as natural resources, and the spaces in which they are found are defined and managed—often as eco-frontiers or enclaves, with governance structures that sustain their management—in ways that are not an outcome of local concerns, priorities, and decision-making (Ramutsindela et al., 2020).

Indigenous Protected Areas

New initiatives in the Arctic are increasingly pushing for the co-production of environmental knowledge, recognising the value of integrating Indigenous knowledge systems with scientific research (e.g., Breton-Honeyman et al., 2021; Hauser et al., 2023). The need to ensure the inclusion of Indigenous and local communities in High Arctic conservation initiatives acquires added urgency given the speed of climate change and resource development projects being scoped out in northern Greenland, as well as in neighbouring Arctic regions. Collaborative conservation and management initiatives, such as those led by the Inuit Circumpolar Council (ICC) and various community-based monitoring programmes, aim to blend Indigenous experiential knowledge with scientific methodologies. Such initiatives acknowledge that Indigenous peoples possess a deep understanding of their surroundings, informed by generations of living in and observing Arctic ecosystems and fostering mutual respect and trust, leading to more effective and inclusive environmental governance. For instance, the Pikialasorsuaq Commission, a joint ICC initiative between Inuit communities in Canada and Greenland, seeks to manage the North Water polynya based on both scientific data and Indigenous knowledge, ensuring that conservation efforts are culturally relevant and locally supported (ICC, 2017).

Largely ice-free throughout the year, the North Water polynya is one of the most productive ecosystems of the planetary ocean. The North Water was named as such by European whalers in the nineteenth century, who went there in search of the Greenland right whale and other marine mammals (Hastrup et al., 2018), but it is called Pikialasorsuaq in Greenlandic, and Sarvarjuaq in the Inuktitut dialect of northern Nunavut. Both mean 'great upwelling'. It sustains key Arctic species, including Arctic cod, seabirds, walrus, narwhal, beluga, polar bear, and several seal species, all of which are essential to the hunting and fishing economies of the Inuit communities in the region, in both Greenland and Nunavut. Yet, like many places, it is vulnerable to climate change (Ribeiro et al., 2021). Its ecological integrity is also threatened by resource development and increased shipping. For example, the only way of transporting ilmenite from a recently opened mine operated by Dundas Titanium—a subsidiary of the UK's Bluejay Mining—at Moriusaq in Northwest Greenland is by ship through a stretch of the polynya. Scientists and conservationists are concerned about these ecosystem changes to the North Water and the increased human commercial activity in the region. So too, as one would imagine they would be, are the people who depend on it.

The World Conservation Union has proposed the North Water as a candidate for a UNESCO Natural Marine World Heritage Site. Research on the vulnerability of the polynya ecosystem by conservation organisations highlights a focus on marine habitats, while their support of scientific work on the habitat of polar bears and ice, as well as whales, again foregrounds the importance they place on charismatic megafauna when promoting the need for environmental protection. Central to the work of ICC's Pikialasorsuaq Commission—to which WWF and Oceans North also contribute—is the argument that Inuit who live in the communities in the Pikialasorsuaq region are the ones who are best placed to set and lead a research agenda, to monitor change, and to establish their own realistic regulations for hunting and wildlife management that would allow for sustainable livelihoods and communities.

Collaborative decision-making processes and co-management arrangements between researchers, government agencies, and Inuit communities in some parts of the Arctic have been evolving since the 1970s (e.g., Danielsen et al., 2020). These processes empower Northern communities and those who use and depend on animals and fish. The Nunavut Land Claims Agreement, for example, sets out principles of conservation and Inuit harvesting rights. Inuit communities and organisations work in collaboration with the appropriate territorial and federal government departments towards co-management systems that integrate science and Inuit Qaujimajatuqangit (IQ). IQ is often translated to mean Inuit traditional knowledge, but more accurately it means "what Inuit have always known to be true" and captures both Inuit epistemology and ontology. Moreover, international organisations and research institutions are increasingly recognising the importance of Indigenous participation in Arctic research. Programmes like those under the auspices of the Arctic Council's Sustainable Development Working Group promote the inclusion of Indigenous knowledge in environmental assessments and policy making. These initiatives emphasise that combining Indigenous insights with scientific research can enhance understanding of climate change impacts and lead to more sustainable adaptation strategies (Johnson et al., 2016). By valuing and incorporating

Indigenous perspectives, these co-production efforts help to decolonise environmental research and ensure that policies are more attuned to the needs and realities of Arctic communities.

In Canada, the federal government supports initiatives for Indigenous Protected Areas in the Arctic—most notably, in new marine protected sites such as the Tallurutiup Imanga/ Lancaster Sound National Marine Conservation Area, in the northern part of Nunavut, where federal and territorial agencies work with local Inuit communities and organisations to conserve biodiversity. A similar initiative—Tuvaijuittuq Marine Protected Area—is underway north and east of Ellesmere Island, Axel Heiberg Island, and other islands in Canada's Arctic Archipelago. There is also state and regional level government recognition of the vital importance of the North Water ecosystem and support for a similarly managed conservation area of Pikialasorsuaq/ Sarvarjuaq that would be cross-boundary in nature, connecting Inuit communities in Nunavut and Greenland, and allowing for collaborative governance between the Canadian state and the Kingdom of Denmark, the territorial government of Nunavut, and the self-rule government of Greenland. The development of Indigenous Protected Areas (IPAs) is gaining momentum as a crucial strategy for biodiversity conservation and the protection of Indigenous rights. IPAs are regions where Indigenous peoples have the authority to manage and protect their lands according to their cultural values, traditional knowledge, and governance systems. In the Arctic, IPAs like the Tallurutiup Imanga National Marine Conservation Area exemplify how Indigenous-led conservation can safeguard both ecological and cultural landscapes. This is intended not only to conserve biodiversity but also reinforce Indigenous sovereignty and self-determination, providing a framework for sustainable management that aligns with the cultural and spiritual values of Indigenous communities.

Globally, the establishment of IPAs is recognised as a powerful tool for achieving conservation goals while respecting Indigenous rights and centring Indigenous voices. Studies have shown that IPAs can be as effectiveas conventional protected areas in preserving biodiversity, if not more so (Garnett et al., 2018; Mansuy et al., 2023). The success of IPAs hinges on the genuine involvement of Indigenous peoples in decision-making processes, as well as their leadership in driving those processes. The expansion of IPAs worldwide underscores the importance of inclusive and participatory conservation models that recognise and elevate Indigenous stewardship, promoting both ecological resilience and social justice. Furthermore, the development and implementation of IPAs aligns with the principles of decolonising the SDGs by empowering Indigenous communities to manage and protect their waters and lands. This approach supports Goal 15 (Life on Land) by promoting biodiversity conservation through Indigenous stewardship, and it also contributes to Goal 16 (Peace, Justice, and Strong Institutions) by reinforcing Indigenous rights and governance systems.

Concluding Thoughts

Contemporary scientific research on the effects of climate change on the world's northern latitudes and recent works on global ecological crisis have unsettled public imaginaries about the nature and seemingly enduring presence of the world's icy places. The Arctic today exemplifies the broader environmental crises facing the planet, and the visual and narrative representation of the region underscores the urgency of addressing climate change. It serves as a critical indicator of the planet's health and the consequences of environmental ruination.

Decolonising conservation necessitates an understanding of the deep and enduring relationality Indigenous communities have with their environment, and the lived expressions of how land, water, and non-human beings are seen not merely as resources to be extracted or preserved but as interconnected entities with inherent rights and agency (cf. Escobar et al., 2024). This contrasts sharply with conventional conservation models, which often isolate human activity from protected areas, imposing strict boundaries between nature and culture. By advocating for Indigenous Protected Areas, Indigenous peoples assert their role not just as stakeholders but as rights holders and primary stewards of biodiversity (Garnett et al., 2018).

In practice, this requires a profound shift in governance models from exclusionary conservation—such as the creation of national parks or moratoria or other restrictions placed on hunting and fishing—to inclusive frameworks that prioritise Indigenous leadership. While global initiatives such as the SDGs and the Convention on Biological Diversity do acknowledge the importance of Indigenous knowledge in biodiversity conservation, these frameworks often fall short of recognising the full scope of Indigenous sovereignty, instead framing Indigenous participation as supplementary to state-led conservation efforts.

Done effectively, collaborative and community-based resource management broadens the understanding of the need to think of ontological spaces in conservation, enhance collaboration, and centre the knowledge, experience, perspectives, and work of those who have previously been excluded. Ways of knowing the environment are multiple and varied, but this does not mean they have to be mutually exclusive. Dialogue that brings these diverse perspectives together is a necessary step for collaborative efforts towards the co-production of the knowledge needed to address and deal with climate change and the many other local, regional, and global challenges that intersect with it.

References

Adams, P. (2014). *Last ice area Greenland and Canada*. Geoscience Resource Development Report. WWF. Retrieved from https://arcticwwf.org/newsroom/publications/last-ice-area-greenland-and-canada/

Adams, W., & Mulligan, M. (2003). *Decolonizing nature: Strategies for conservation in a post-colonial era*. Routledge.

Appadurai, A. (1990). Disjuncture and difference in the global economy. *Theory, Culture & Society, 7*, 295–310.
Asiyanbi, A. D., Ogar, E., & Akintoye, O. A. (2019). Complexities and surprises in local resistance to neoliberal conservation: Multiple environmentalities, technologies of the self and the poststructural geography of local engagement with REDD+. *Political Geography, 69*, 128–138.
Bennett, N. J., Govan, H., & Sattrefield, T. (2015). Ocean grabbing. *Marine Policy, 57*, 61–68.
Biermann, C., & Anderson, R. M. (2017). Conservation, biopolitics and the governance of life and death. *Geography Compass, 11*(10), e12329. https://doi.org/10.1111/gec3.12329
Blaser, M. (2016). Is another cosmopolitics possible? *Cultural Anthropology, 31*(4), 545–570.
Breton-Honeyman, K., Huntington, H. P., Basterfield, M., Campbell, K., Dicker, J., Gray, T., Jakobsen, A. E., Jean-Gagnon, F., Lee, D., Laing, R., Loseto, L., McCarney, P., Noksana, J., Jr., Palliser, T., Ruben, L., Tartak, C., Townley, J., & Zdor, E. (2021). Beluga whale stewardship and collaborative research practices among indigenous peoples in the Arctic. *Polar Research, 40*(S1). https://doi.org/10.33265/polar.v40.5522
Büscher, B., & Fletcher, R. (2020). *The conservation revolution: Radical ideas for saving nature beyond the Anthropocene*. Verso.
Buschman, V. O. & Sudlovenick, E. (2023). Indigenous-led conservation in the Arctic supports global conversation practices. *Arctic Science, 9*(3): https://doi.org/10.1139/as-2022-0025
Danielsen, F., Johnson, N., Lee, O., Fidel, M., Iversen, L., Poulsen, M. K., Eicken, H., Albin, A., Hansen, S. G., Pulsifer, P. L., & Thorne, P. (2020). *Community-based monitoring in the Arctic*. University of Alaska Press.
Davis, H., & Todd, Z. (2017). On the importance of a date, or decolonizing the Anthropocene. *ACME: An International Journal for Critical Geographies, 16*(4), 761–780.
Dawson, A., Longo, F., & Survival International. (2023a). *Decolonize conservation: Global voices for indigenous self-determination, land, and a world in common*. Common Notions.
Dawson, N., Coolsæt, B., Bhardwaj, A., Booker, F., Brown, D., Lliso, B., Loos, J., Martin, A., Oliva, M., Pascual, U., Sherpa, P., & Worsdsell, T. (2023b). Is it just conservation? A typology of indigenous peoples' and local communities' roles in conserving biodiversity. *One Earth, 7*. https://doi.org/10.1016/j.oneear.2024.05.001
DePuy, W., Weger, J., Foster, K., Bonanno, A. M., Kumar, S., Lear, K., Basilio, R., & German, L. (2022). Environmental governance: Broadening ontological spaces for a more livable world. *Environment and Planning E: Nature and Space, 5*(2), 947–975.
Escobar, A., Osterweil, M., & Sharma, K. (2024). *Relationality: An emergent politics of life beyond the human*. Bloomsbury.
Foss, Ø., Maton, J., Moholdt, G., Schmidt, L. S., Sutherland, D. A., Fer, I., Nilsen, F., Kohler, J., & Sundfjord, A. (2024). Ocean warming drives immediate mass loss from calving glaciers in the high arctic. *Nature Communications, 15*, 10460. https://doi.org/10.1038/s41467-024-54825-7
Garnett, S. T., Burgess, N. D., Fa, J. E., Fernández-Llamazares, Á., Molnár, Z., Robinson, C. J., Watson, J. E., Zander, K. K., Austin, B., Brondizio, E. S., & Collier,

N. F. (2018). A spatial overview of the global importance of indigenous lands for conservation. *Nature Sustainability, 1*, 369–374.

Hastrup, K., Mosbech, A., & Grønnow, B. (2018). Introducing the north water: Histories of exploration, ice dynamics, living resources, and human settlement in the Thule region. *Ambio, 47*(Suppl. 2), 162–174.

Hauser, D. D. W., Glenn, R. T., Lindley, E. D., Pikok, K. K., Heeringa, K., Jones, J., Adams, B., Leavitt, J. M., Omnik, G. N., Schaeffer, R., SimsKayotuk, C., Sparrow, E., Ravelo, A. M., Lee, O., & Eicken, H. (2023). Nunaaqqit Savaqatigivlugich—Working with communities: Evolving collaborations around an Alaska Arctic observatory and knowledge hub. *Arctic Science, 9*(3), 635–656. https://doi.org/10.1139/as-2022-0044

Heide-Jørgensen, M. P., Garde, E., Hansen, R. G., Tervo, O. M., Sinding, M.-H. S., Witting, L., Marcoux, M., Watt, C., Kovacs, K. M., & Reeves, R. R. (2020). Narwhals require targeted conservation. *Letters, Science, 370*(6615), 416. https://doi.org/10.1126/science.abe7105

Heise, U. K. (2016). *Imagining extinction: The cultural meanings of endangered species*. The University of Chicago Press.

Hill, J. (2009). *White horizon: The Arctic in the nineteenth-century British imagination*. State University of New York Press.

Hodgetts, T. (2017). Wildlife conservation, multiple biopolitics and animal subjectification: Three mammals' tales. *Geoforum, 79*, 17–25.

Igoe, J. (2017). *The nature of spectacle: On images, money, and conserving capitalism*. University of Arizona Press.

Inuit Circumpolar Council. (2017). *People of the ice bridge: The future of the Pikialasorsuaq*. Report of the Pikialasorsuaq Commission. Inuit Circumpolar Conference.

Jahn, A., Holland, M. M., & Kay, J. E. (2024). Projections of an ice-free Arctic Ocean. *Nature Reviews Earth and Environment, 5*, 164–176. https://doi.org/10.1038/s43017-023-00515-9

Johnson, N., Behe, C., Danielsen, F., Krummel, E-M., Nickels, S., Pulsifer, P. L. (2016). Community-Based monitoring and indigenous knowledge in a changing arctic: A review for the sustaining arctic observing networks. *Sustain Arctic Observing Network Task, 9*. https://iccalaska.org/wp-icc/wp-content/uploads/2016/05/Community-Based-Monitoring-and-Indigenous-Knowledge-in-a-Changing-Arctic_web.pdf

Mansuy, N., Staley, D., Alook, S., Parlee, B., Thomson, A., Billie Littlechild, D., Munson, M., & Didzena, F. (2023). Indigenous protected and conserved areas (IPCAs): Canada's new path forward for biological and cultural conservation and indigenous well-being. *FACETS, 8*, 1–16. https://doi.org/10.1139/facets-2022-0118

McCorristine, S. (2018). *The spectral Arctic: A history of dreams and ghosts in polar exploration*. UCL Press.

McCorristine, S., & Adams, W. M. (2020). Ghost species: Spectral geographies of biodiversity conservation. *Cultural Geographies, 27*(1), 101–115.

Millan, R., Jager, E., Mouginot, J., Wood, M. H., Larsen, S. H., Mathiot, P., Jourdain, N. C., & Bjørk, A. (2023). Rapid disintegration and weakening of ice shelves in North Greenland. *Nature Communications, 14*, 6914. https://doi.org/10.1038/s41467-023-42198-2

Newton, R., Pfirman, S., Tremblay, L. B., & DeRepentigny, P. (2021). Defining the "ice shed" of the arctic ocean's last ice area and its future evolution. *Earth's Future, 9*, e2021EF001988. https://doi.org/10.1029/2021EF001988

Notz, D., & Stroeve, J. (2018). The trajectory towards a seasonally ice-free Arctic Ocean. *Current Climate Change Reports, 4*, 407–416. https://doi.org/10.1007/s40641-018-0113-2

Nuttall, M. (2017). *Climate, society and subsurface politics in Greenland: Under the great ice*. Routledge.

Nuttall, M. (2019). Icy, watery, liquescent: Sensing and feeling climate change on Northwest Greenland's coast. *Journal of Northern Studies, 14*(2), 71–91.

Nuttall, M. (2024). Understanding Arctic melt: Reflections on collaborative interdisciplinary research. In S. A. Crate & M. Nuttall (Eds.), *Anthropology and climate change: From transformations to worldmaking* (pp. 77–92). Routledge.

Potter, R. A. (2007). *Arctic spectacles: The frozen north in visual culture, 1818–1875*. University of Washington Press.

Ramutsindela, M., Guyot, S., Boillat, S., Giraut, F., & Bottazi, P. (2020). The geopolitics of protected areas. *Geopolitics, 25*(1), 240–266.

Ribeiro, S., Limoges, A., Massé, G., Johannsen, K. L., Colgan, W., Weckström, K., Jackson, R., Georgiadis, E., Mikkelsen, N., Kuijpers, A., Olsen, J., Olsen, S. M., Nissen, M., Strunk, A., Wetterich, S., Syväranta, J., Henderson, A. C. G., Mackay, H., Taipale, S., Jeppesen, E., Larsen, N. K., Crosta, X., Giraudeau, J., Nuttall, M., Grønnow, B., Mosbech, A., & Davidson, T. A. (2021). Vulnerability of the north water ecosystem to climate change. *Nature Communications, 12*, 4475. https://doi.org/10.1038/s41467-021-24742-0

Schnegg, M. (2021). Ontologies of climate change. *American Ethnologist, 48*(3), 260–273.

Ulloa, A. (2017). Thee geopolitics of carbonized nature and the zero carbon citizen. *South Atlantic Quarterly, 116*(1), 111–120.

Ulloa, A. (2019). Indigenous knowledge regarding climate in Colombia: Articulations and complementarities among different knowledges. In G. Feola, H. Geoghegan, & A. Arnall (Eds.), *Climate and culture: Multidisciplinary perspectives on a warming world*. Cambridge University Press.

Villas Boas, G. H., Marçal, M., & Brierley, G. (2022). An ethnogeomorphic case study of conservation practices in Southeast Brazil. *Human Ecology, 50*, 651–664. https://doi.org/10.1007/s10745-022-00345-1

West, P. (2006). *Conservation is our government now: The politics of ecology in Papua New Guinea*. Duke University Press.

Williams, G. (2009). *Arctic labyrinth: The quest for the Northwest Passage*. Viking Canada.

CHAPTER 4

Decolonisation and Green Energy in Sápmi: A Trade-off Analysis of the Sustainability Development Goals

Reetta Toivanen

Introduction

The Arctic is warming at an alarming rate, leading to the melting of ice caps, rising sea levels, and alterations in wildlife habitats (Kreplin et al., 2021; Thomas (ed.), 2020). These environmental changes can exacerbate existing vulnerabilities and create new challenges, particularly for Indigenous peoples who have lived sustainably in these regions for hundreds of years. As traditional hunting grounds and ways of life are disrupted (Raitio et al., 2020), there is a pressing need to understand the multifaceted impacts of climate change on both the environment and human societies. The tensions between climate action and the protection of human rights are particularly pronounced in the Arctic (Toivanen & Cambou, 2021). Indigenous peoples often bear the brunt of climate change, facing displacement, loss of access to traditional lands, and threats to their cultural practices (Nuttall, 2021; see also this book).

Critical social scientists argue that current frameworks, such as Agenda 2030 and the Sustainable Development Goals (SDGs), may overlook alternative models of development that could be more appropriate for the unique contexts of the Arctic—including the people living there (Stepien et al., 2014). The current frameworks often emphasise economic growth and technological solutions, which may not align with the values and practices of Indigenous and

R. Toivanen (✉)
Department of Cultures, University of Helsinki, Helsinki, Finland
e-mail: reetta.toivanen@helsinki.fi

R. Toivanen et al. (eds.), *Decolonizing the Sustainable Development Goals*,
https://doi.org/10.1007/978-3-032-14263-4_4

local communities (Haley, 2023). The danger that climate-related policies do not adequately incorporate Indigenous voices and rights already exists (Natcher & Koivurova, 2021). Therefore, there is a call for decolonising approaches that are more holistic, recognising the interconnectedness of social, environmental, and cultural factors while also paying due attention to questions of power.

This chapter focuses on decolonising of the SDGs by providing a trade-off analysis of the different sustainability goals and their respective indicators, as defined by the United Nations (UN) (UN, 2015). The question is: How does the conflict between the interests of Sámi reindeer herders and the development of wind power and other climate change mitigation measures find expression in the SDGs and their interactions? Thus, how well do the SDGs recognise the life realities and livelihoods of Indigenous and other local populations residing in the Arctic? The chapter contributes to an understanding of why the talk about what should be done in order to "reach" the United Nations Sustainability Goals is harmful for a large part of the world's inhabitants. This is especially true for marginalised and vulnerable groups, such as Indigenous peoples, who have been sidelined in the SDG decision-making process.

Setting the Neoliberal Scene

Let us begin with an extract of a discussion with three reindeer herders in a small town in the Arctic in Finland.

> *Person 1:* It is a fight that we have had since I have been an adult, going to court against Sokli (mine in Savukoski). They always say that the reindeer-herding collaborative will be heard.
>
> *Person 2:* Yes yes! Indeed, you are heard but not listened to!
>
> *Person 1 and 3:* (approving laughter)
>
> *Person 3:* They have, so one could say, not given us the opportunity to influence decisions…
>
> *Reetta:* What would it mean to your reindeer-herding collaborative if the mine would come here?
>
> *Person 1:* It would…
>
> *Person 2:* It would go in two halves…
>
> *Person 1:* Yes, so it would…
>
> *Person 2:* The whole reindeer herd would be divided into two and the mine would be in the middle.
>
> *Person 1:* And there would be no summer pasture.
>
> *Person 3:* Even though the area for the actual mine is small, the multiplication impacts, roads and what not would divide our reindeers into two and we could not work anymore.

Here, people from a small town near the border of the officially recognised Sámi home territory[1] are talking with me and my colleague about the impacts

[1] The official territory of Sámi homeland is defined in the Act on the Sámi Parliament 974/1995. Numerous Sámi families live just outside the border of the official homeland.

of the Sokli mine.[2] We hear them joking about their possibility (or lack thereof) to take part in any negotiation about their livelihoods. We also hear that the mine would mean the end of their lifestyle as reindeer herders. Since the mine is not located in the officially recognised homeland of the Indigenous peoples in Finland, the Sámi Parliament (the representative of the Indigenous community) has no say about the mine either. Nor can the UN-recognised principles of Free, Prior, and Informed Consent (FPIC)[3] be called upon here, since the municipality is outside the official homeland of Sámi peoples.

Traditional Sámi livelihoods include reindeer herding, fishing, hunting, handicrafts, and gathering (Jaakkola et al., 2018). Reindeer herding is economically vital and central to Sámi culture, though only a minority of Sámi engage in it (Normann, 2021). The Sámi, particularly reindeer herders, are vulnerable to environmental changes, reporting shifts in seasons and vegetation, extreme weather events, and reduced grazing lands (Furberg et al., 2011; Nysten-Haarala et al., 2021). Additionally, industrial activities like mining and wind parks limit grazing areas and disturb reindeer (Skarin & Åhman, 2014; Österlin & Raitio, 2020).

Scholars note that in negotiations concerning northern Finland (and similarly in Sweden; see Cambou, 2020), reindeer herding and other Indigenous livelihoods are often viewed as just another competing business. This perspective persists despite the UN Human Rights Committee's recognition over two decades ago in the Länsman vs Finland case (1997) that Indigenous livelihoods are vital for preserving culture and language, not just economic interests. Yet, the Finnish state continues to prioritise natural resource economies that depend on industries like logging, mining, or oil extraction over more diversified economies that include the cultural value of reindeer herding (Toivanen, 2019).

Often portrayed as empty and pristine (Toivanen, 2019), the Arctic serves as a backdrop for neoliberal policies, as decision-makers and corporations view it as a resource frontier for unrestrained extraction. Many are surprised to learn that the Arctic is home to human communities and wildlife, and that the landscape has significant cultural value (Toivanen & Fabritius, 2020). In fact, the Arctic is marked by environmental and social diversity, facing such challenges as rapid warming, demographic shifts, and geopolitical tensions, which have intensified since Finland and Sweden joined NATO (Nilsson & Larsen, 2020). Though claimed by eight states, the Arctic functions as an international region, exemplified by the Arctic Council established in 1996 (Keskitalo, 2007). However, the Russian war in Ukraine has disrupted collaboration within the Council, as tensions between member states have challenged its ability to

[2] https://www.mineralsgroup.fi/business/sokli.html (last accessed 19 November 2024).

[3] Free, Prior, and Informed Consent (FPIC) is a specific right granted to Indigenous peoples recognised in the UN Declaration on the Rights of Indigenous Peoples (UNDRIP), which aligns with their right to self-determination. The FPIC allows Indigenous peoples the ability to provide, withhold, or withdraw consent at any point regarding projects impacting their territories. For more see https://www.fao.org/indigenous-peoples/our-pillars/fpic/en/ (last accessed 1 October 2024).

function as a unified platform for addressing Arctic issues. Exploring global strategies with an Arctic perspective is crucial for understanding the region's unique conditions.

The 17 Sustainable Development Goals and their 232 indicators have been criticised for inadequately reflecting the economic and cultural realities essential for sustainable development in the Arctic (Sköld et al., 2018; Nilsson & Larsen, 2020). Indigenous peoples, making up about 10% of the Arctic population, face severe climate change impacts on marine and freshwater ecosystems that threaten their livelihoods (Nuttall, 2018). Indigenous perspectives are largely absent from the SDGs, mentioned only in Targets 2.3 and 4.5, where they are categorised as a vulnerable group, overlooking their specific needs and rights (Yap & Watene, 2019). This lack of representation has also been critiqued in the Arctic context (Degai & Petrov, 2021).

Indigenous people in the Arctic are not only affected by climate change and its environmental impacts but also by policies and measures that aim to combat climate change (Cambou, 2020; Normann, 2021). An example of this is the development and expansion of wind power in northern Scandinavia and its impacts on the activities of the local Indigenous communities (see also von der Pütten in Chap. 5 and Elle Eriksson & May-Britt Öhman in Chap. 9). A conflict arises between protecting Indigenous culture and practices, on one hand, and promoting renewable energy sources, on the other.

The Concept of Decolonisation

Decolonialism in the Arctic refers to efforts to address and undo the historical and ongoing impacts of colonialism on the Indigenous peoples and ecosystems of the region. Colonialism in the Arctic has largely involved the domination, exploitation, and control of the land, resources, and cultures of Indigenous Arctic populations (such as the Sámi, Inuit, and other Native groups) by external powers, mainly European nations, and later, the United States and Canada. This colonial history has had significant consequences, including loss of land, suppression of languages and cultures, forced relocations, and environmental degradation due to resource extraction. As Tuhiwai Smith (2012, p. 40) notes, a decolonising approach requires actively promoting and validating alternative narratives and perspectives. She highlights the exclusion of non-Western viewpoints from academic discussions, which marginalises knowledge and silences colonised communities.

In this research, decolonialism emphasises reversing colonial legacies by restoring power, rights, and agency to the Indigenous peoples of the Arctic. This includes reclaiming land rights, promoting Indigenous governance, and valuing traditional knowledge often dismissed by colonial powers (Tuck & McKenzie, 2014; Simpson, 2020). Decolonialism also seeks to preserve Indigenous languages, practices, and identities while ensuring that development projects like resource extraction prioritise Indigenous needs and protect the fragile Arctic environment.

Methods

The main dataset for the trade-off analysis in this chapter is based on desktop research collected in a project funded by the Arctic Avenue.[4] A trade-off analysis between different Sustainable Development Goals (SDGs) evaluates how achieving progress in one goal may positively or negatively impact others. It identifies synergies (win-win situations) and trade-offs (conflicting outcomes) to balance competing priorities, ensuring holistic and efficient decision-making for sustainable development (Hegre et al., 2020). Selected in the project were those SDG indicators directly affecting the nature substance economies in the Arctic area. The selected targets are directly related to an earlier identified tension between reindeer herders and wind energy projects, so they either represent Indigenous activities or relate to the implementation and planning of wind power projects.

The other dataset consists of thematic group interviews carried out by the author and sometimes together with other colleagues[5] in Finnish Lapland since 2011 with a specific focus on wellbeing in historically multilingual and multicultural municipalities. As it was clear at the outset that extractive industries create a conflict with natural subsistence economies, it was nevertheless important to discuss both the negative and positive impacts of industries in the Arctic. As positive impacts, salaried jobs were always mentioned, in addition to new cultural influences, modernisation of towns, and educational opportunities. Negative impacts include the end of natural substance economies on the whole and, with them, the end of connectedness to nature, considered the home of the Sámi languages and cultures. This data will be used to inform and substantiate the trade-off analysis of the SDG indicators.

Ethical Considerations and Researcher's Positionality

The research for this chapter was conducted in strict adherence to the Finnish Ethical Board's regulations (TENK, 2019), as well as the emerging Ethical Guidelines of Sámi Research (approved in 2024; see Heikkilä et al., 2024). The interviewed participants received a letter explaining the research goals and questions and the use and storage of the interview material. Recorded oral permission for the interview was obtained. The interviewed persons also had the option to later withdraw from the research. The author, who is an ethnic Finn, carried out many of the interviews together with or in the presence of Sámi academics. The interviews and ethnographic material for this chapter presents just a small part of the complete dataset, which also deals with other issues and questions relevant to the lives of Sámi and local residents in the Arctic.

[4] Arctic Avenue fund with Zahra Kalantari, Department of Sustainable Development, Environmental Science and Engineering (SEED), KTH Royal Institute of Technology, Stockholm, Sweden; Carla Sofia Santos Ferreira and Hanna Kreplin (University of Stockholm); and Reetta Toivanen and Anna Varfolomeeva (University of Helsinki), 2021.

[5] Part of the research was carried out in collaboration with Janne Saarikivi and Sanna Valkonen.

What Is a Green or Clean Transition and What Does It Mean Locally?

Synergies Explained

Synergy and trade-off analyses can be performed using quantitative or qualitative methods. The former is based on statistical analyses using indicator data. A correlation between certain indicators would then result in a synergy between the targets. Examples for such analyses are Pradhan et al. (2017) and Hegre et al. (2020). Focusing on the Arctic complicates such quantitative methods, as indicator data is often collected per country, whereas the Arctic spans a region with the northernmost parts of different countries. It therefore becomes difficult to collect data specifically for the Arctic. Qualitative analyses are based on questionnaires, interviews, and literature reviews.

In total, 14 targets were selected. These are presented together with their indicators in Fig. 4.2 at the end of this chapter. It includes targets aimed at the protection of ecosystems and reducing the degradation of natural habitats, such as Targets 6.6, 14.2, 15.1, and 15.5. Other targets represent the rights of Indigenous people regarding access to economic and natural resources (Target 1.4), as well as their political inclusion and legal discrimination (Target 10.2, on empowerment) and reducing inequalities of outcome (10.3), which becomes relevant in permit processes of wind power projects. Target 11.4 was included as it is the only target aimed at the safeguarding of cultural heritage, which can also be interpreted in regard to the protection of Indigenous Sámi culture. Climate change is relevant for both Indigenous activities and wind power projects, which led to the inclusion of Target 13.3. The planning and implementation of wind power projects is represented in Targets 7.1, 7.2, and 7.3 from SDG 7 (Affordable and Clean Energy). In addition, Target 8.1 represents economic growth; it was included as clean energy projects can lead to economic growth in an area with little industry. Similarly, Target 9.1 aims to improve infrastructure, which is also directly related to wind power projects in northern Fennoscandia.

The literature review was performed with a focus on interactions between the above targets with a broad approach, in order to identify certain synergies and trade-offs. During the analysis, certain targets were found to be of the same overall theme or topic and could therefore be summarised as six clusters (see Fig. 4.2). These include Indigenous culture and economy (Targets 1.4 and 11.4), political inclusion and justice (Targets 10.2 and 10.), nature conservation (Targets 6.6, 14.2, 15.1, and 15.5), economic growth and infrastructure (Targets 8.1 and 9.1), clean energy (Targets 7.1, 7.2, and 7.3), and climate change adaptation and mitigation (Target 13.3). Within these clusters, the targets positively affect one another, creating synergies. The identified synergies and trade-offs are shown in Fig. 4.1 by green and red arrows, respectively. Figure 4.1 shows the synergies and trade-offs between all individual targets in a table format, as suggested by Weitz et al. (2018). So far, 44 synergies and 18 trade-offs, as well as four mixed interactions (representing both synergy and

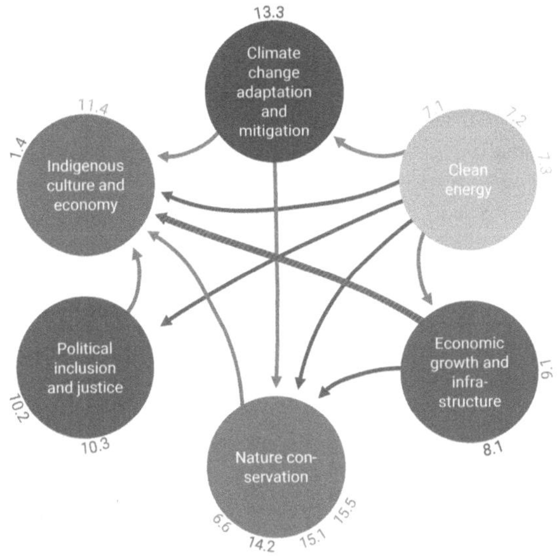

Fig. 4.1 Synergies and trade-offs between SDG targets (created by Hanna Kreplin)

trade-off) of 182 possible links, could be determined based on a preliminary literature analysis (see Fig. 4.2).

Clean Energy Versus Political Inclusion and Justice

The permit processes of wind power projects are often determined and affected by a multitude of laws and institutions, creating complications and impeding inclusion of the affected actors, such as the Indigenous Sámi people. As planning processes are fragmented and top-down, the active participation of Sámi in this phase is limited (Larsen & Raitio, 2019; Bjärstig et al., 2020). In Sweden, compared to Finland where the right to be heard is established by the Law on the Sámi Parliament (974/1995), there is no legal requirement to consult with Sámi as Indigenous people. The participation of affected authorities and individuals, including ethnic minorities, is stipulated by laws, but municipalities can freely decide how participation with different stakeholders will be executed (Cambou, 2020). Often, Sámi only receive or exchange information, and they are limited in their ability to influence such planning processes (Larsen et al., 2017). Sámi rights to the land exist in the form of usage rights, in contrast to ownership, which further diminishes their participation in such planning processes (Toivanen, 2015). This impedes and exacerbates the social, economic, and political inclusion of the Sámi, represented by Target 10.2 (political participation). Therefore, a trade-off can be determined between Targets 7.2 (renewable energy) and 10.2 (political participation).

The permitting processes for wind power projects raise issues of justice and discrimination. Cambou (2020) details how lawsuits in Sweden often disregard Sámi interests, putting their procedural rights at risk. While reindeer husbandry is generally protected from interference, clear guidelines for balancing the interests of reindeer herders and wind power investments are lacking. In areas suitable for both, the focus is solely on promoting "sustainable development" (Cambou, 2020). Previous court rulings suggest that permits are denied if a wind project would threaten reindeer herding; however, if such harm cannot be proven, courts tend to prioritise renewable energy projects (Cambou, 2020).

In terms of wind power construction in the whole of Sweden, 7% of permits have been rejected due to interference with reindeer husbandry; this represents 10% of turbines not being installed (Cambou, 2020). Generally, there is a tendency for more permits to be denied in the south of Sweden than the north, as the south is more densely populated and the north is therefore more attractive for large-scale projects (Cambou, 2020; see also Elle Eriksson & May-Britt Öhman in Chap. 9 in this book). This demonstrates the distributional injustice that Sámi are facing (Liljenfeldt & Pettersson, 2017). Similar processes and court decisions in the context of reindeer herding and wind power projects have also been reported in Norway (Normann, 2021). Such issues of injustice can be interpreted as discriminatory practices, referring to Target 10.3, which might therefore result in a trade-off between Targets 7.2 (renewable energy) and 10.3 (political inclusion).

Clean Energy and Indigenous Economy and Culture

In previous lawsuits, courts have tended to consider reindeer herding as an economic business, neglecting the important traditional, cultural, and spiritual aspects of reindeer husbandry for the Sámi (Lawrence, 2014; Normann, 2021). The conflict between wind power projects and reindeer herding is often characterised as a clash between two competing industries. Given the economic advantages of wind power, court rulings tend to favour the approval of those projects, with reindeer herders typically receiving compensation to mitigate their financial losses (Normann, 2021). Referring to Target 1.4, which states that "all men and woman [shall] [...] have equal rights to economic resources", the construction of wind turbines diminishes the Sámi's access to economic resources, such as reindeer. Similarly, the Sámi's lack of access to "ownership and control over land" is also exacerbated by wind power projects, since the Sámi only hold land-use rights, not land ownership. Therefore, a trade-off between Targets 7.2 (renewable energy) and 1.4 (equal rights to economic resources) can be determined. The economic aspect of the conflict can also be found in Target 10.2, mentioned above.

More importantly, the consideration of reindeer herding solely as a business disregards the Indigenous culture and spirituality of the Sámi, reflecting the context of colonisation and oppression of the Sámi and their cultural practices (Lawrence, 2014). Understanding the role and importance of reindeer for the survival of Sámi culture, even though most Sámi are not personally involved in reindeer herding, is therefore crucial for investigating the impact of wind power projects on Sámi traditions and livelihoods.

The relationship between Sámi and reindeer has evolved over centuries. Initially, Sámi livelihoods relied on fishing, berry picking, and hunting animals like wild deer, bears, and beavers (Sara, 2009; Bergman et al., 2013). Small-scale reindeer herding was also practised, using reindeer for transport, milk, and meat (Müller-Wille et al., 2006). Sámi social organisation, known as *Siida*, united families based on locality (Virtanen & Seurujärvi-Kari 2019). During spring and summer, families moved according to resource availability, while winter saw gatherings in villages for trade and contact with merchants and tax collectors (Sara, 2009). With the decline of wild deer in the late sixteenth and early seventeenth centuries, nomadic reindeer herding became more prevalent, leading to the establishment of distinct herding districts (Sara, 2009; Bergman et al., 2013). This shift transformed Sámi lifestyles, as managing larger herds required full-time dedication (Müller-Wille et al., 2006; Bergman et al., 2013).

The harsh Arctic conditions fostered a deep cultural and spiritual connection between the Sámi and their reindeer, making herding central to their identity and a means to transmit knowledge across generations (Normann, 2021). However, the expansion of southern societies and colonisation threatened Sámi traditions through border closures in Norway (eighteenth century) and Russia (nineteenth century), leading to community divisions and restrictions on reindeer herding, fishing, and hunting. Additionally, colonial practices, particularly in churches and schools, gradually eroded Sámi culture (Salmi & Heino, 2019). Current legislation issues and the insecure legal basis of reindeer herding (Nysten-Haarala et al., 2021), as well as processes disrupting their activities, also harm Sámi culture, causing stress and mental health problems among reindeer herders (Daerga et al., 2008; Kaiser et al., 2010). Activities that affect Sámi practices, such as the development of wind power projects, stand in conflict with the protection and preservation of cultural heritage. This therefore represents a trade-off between Targets 7.2 and 11.4.

Political Inclusion and Indigenous Economic Rights

Since the early modern period, the Sámi have been subject to enduring processes of marginalisation in relation to Nordic state formation and governance (Müller-Wille et al., 2006). Their colonisation included restrictions and limitations on Sámi activities, and the Sámi were subject to many geopolitical changes throughout the centuries (Henriksen, 2008). The discrimination of Sámi further worsened after industrialisation, when the economic activities of Norway, Sweden, Finland, and Russia grew in the north. Sámi were forced to assimilate, and their culture and livelihoods were eroded and marginalised (Selle & Strømsnes, 2010). Political inclusion in the form of active participation and access to institutional processes as well as procedural rights fosters today's Indigenous self-determination (Toivanen, 2019). An important step in this regard was the establishment of the Nordic Sámi Council in 1956 (today the Sámi Council), aimed at promoting and safeguarding Sámi rights and interests (Henriksen, 2008). Eliminating discriminative laws, policies, and practices will further help the Sámi to maintain their livelihoods, traditions, and culture

(Henriksen, 2008; Broderstad, 2014). This therefore represents a synergy between Targets 10.2 and 10.3 (empowerment and non-discrimination) and Targets 1.4 and 11.4 (economic rights and right to cultural heritage).

Clean Energy Versus Economic Growth and Infrastructure

Wind power projects generally promote economic growth (Target 9.1) in remote and rural areas, such as northern Fennoscandia. The biggest impact in this regard is the creation of new jobs as well as increased incomes (Brown et al., 2012; Liljenfeldt, 2013; Aldieri et al., 2020). New jobs evolve mainly during the construction phase, including both jobs directly related to the construction of the wind turbines and indirectly linked to employment in sectors such as public administration, commercial services, hotels, and restaurants (Ejdemo & Söderholm, 2015). This effect is even higher when turbine materials, such as blades and towers, are also produced in the same area (Munday et al., 2011; Ejdemo & Söderholm, 2015). During the operating phase, however, the number of jobs created directly and indirectly by the wind power project is fewer than in the construction phase (Ejdemo & Söderholm, 2015).

Further positive impacts of wind power projects on the local economic development include increased tax revenues, lease payments to landowners, new opportunities for local businesses, and other community benefits (Brown et al., 2012; Liljenfeldt, 2013; Ejdemo & Söderholm, 2015). This therefore represents a synergy between Targets 7.2 and 8.1. Similarly, wind power projects in rural areas have positive impacts on the development of infrastructure (Target 9.1). New roads, power lines, and other structures are required for wind power plants and are therefore built together with the actual turbines (Aitken, 2010; Liljenfeldt, 2013; Ejdemo & Söderholm, 2015). These improvements in infrastructure offer additional community benefits, apart from economic growth (Liljenfeldt, 2013). This hence results in a synergy between clean energy and infrastructure development (i.e. Targets 7.2 and 9.1). However, the Indigenous perspective is missing here.

Economic Growth and Indigenous Economy

During the fifteenth and sixteenth centuries, Norway, Sweden (including Finland), and Russia expanded their territories to the north, spatially restricting the activities of the Sámi to a great extent and colonising their communities (Müller-Wille et al., 2006). In this regard, the Sámi lost land, resources, and authority in northern Fennoscandia to their colonisers (Brännlund & Axelsson, 2011; Lawrence & Åhrén, 2016). The economic development in the north mainly included logging and forestry, as well as mining, and impacted the traditional reindeer husbandry negatively by decreasing grazing areas and restricting the movements of reindeer (Müller-Wille et al., 2006). Similarly, tourism and other recreation land uses, which emerged later in time, have required extensive infrastructure and interfered with reindeer-herding practices

(Müller-Wille et al., 2006). During the twentieth and twenty-first centuries, reindeer herding has undergone drastic changes due to revisions in legislation, hygienic measures for herding and slaughtering, quota systems for grazing districts, and more bureaucratic processes (Heikkinen, 2006). This indicates a negative impact and hence a trade-off between economic growth and infrastructure (Targets 8.1 and 9.1) and Indigenous economy and culture (Targets 1.4 and 11.4).

However, the relationship between economy and culture is more complex. For instance, Sámi reindeer herders have adopted modern technologies and infrastructure, such as snowmobiles, trailer transportation, and GPS localisation, which have helped the herders and facilitated reindeer herding, particularly during the winter months (Williams, 2003). Reindeer herding is now an economically viable business, and herders are entrepreneurs selling their products to a much larger market (Heikkinen, 2006). However, they are under strong pressure, as they must compete with larger, independent, and external companies that determine the price of reindeer meat (Müller-Wille et al., 2006). Tourism has developed as an additional business model for Sámi. It serves as a source of employment and income, and it also allows them to act as cultural ambassadors and teach outsiders about their traditions (Leu et al., 2018). Whether this relationship can be determined as synergy or a trade-off remains to be discussed.

Clean Energy and Non-Humans

Wind power is a crucial renewable energy source for reducing greenhouse gas emissions and mitigating climate change. However, developing wind farms in remote areas affects local environments and ecosystems (Targets 6.6, 14.2, 15.1, and 15.5). Turbines can induce turbulence, altering land surface and air flows, which may impact local weather and increase surface temperatures (Ren et al., 2018; Zhou et al., 2012).

The environmental effects on bird species are significant and have garnered considerable attention. Studies show that collisions with turbines and related structures lead to decreased bird populations and habitat loss, both onshore and offshore (Wang et al., 2015; Wang & Wang, 2015; Beston et al., 2016; May et al., 2021). While wind power causes fewer bird fatalities compared to nuclear and fossil fuel sources (Sovacool, 2013), it also contributes to bat mortality and habitat destruction (Wang & Wang, 2015; Gaultier et al., 2020). Fortunately, recent technological advancements have reduced bird fatalities associated with turbines (Wang & Wang, 2015). Additionally, wind farms generate mechanical and aerodynamic noise pollution, which can disrupt animal communication and contribute to wildlife fatalities (ibid.).

The environmental impact of wind power plants on reindeer has been studied in Scandinavia (Helldin et al., 2012; Skarin & Åhman, 2014; Skarin et al., 2015). Increased noise, human activity, and infrastructure development disrupt reindeer grazing and fragment habitats (Skarin et al., 2015). Wind turbines also cause visual disturbances that particularly affect prey species like reindeer (Skarin & Åhman, 2014; Skarin et al., 2018). The construction of roads and infrastructure for wind power facilitates human access to remote areas for recreation and hunting, potentially leading to habitat loss (Helldin et al., 2012). As new wind farms are planned for remote locations, these environmental impacts may increase (Österlin & Raitio, 2020). Conversely, roads can benefit larger mammals by improving movement and creating new browsing areas (Helldin et al., 2012). The extent of the disturbance may vary, as animals adapt to new structures, and the impacts may depend on the size and number of turbines in a given area (ibid.). Further research is needed to understand the full effects of wind turbines on mammals, particularly reindeer, in northern Fennoscandia. The above-described impacts of wind power plants on wildlife, including habitat destruction, fatalities, and decreases in species population, represent a conflict between the targets of clean energy and nature conservation. This includes both freshwater and coastal ecosystems and hence causes a trade-off between Target 7.2 (renewable energy) and Targets 6.1, 14.2., 15.1, and 15.5 (nature conservation).

Nature Conservation Versus Indigenous Economy and Culture

The Sámi and their traditional reindeer herding rely on healthy ecosystems with sufficient grazing areas (Müller-Wille et al., 2006). Protecting these habitats and biodiversity is vital for sustaining Sámi culture and economy, as reindeer are key economic resources. This also encompasses other important aspects of Sámi livelihoods, such as fishing, hunting, handicrafts, and gathering. Conflicts may arise between the Sámi and Environmental Protection Agencies over nature conservation and the designation of protected areas (Target 15.1 and Indicator 15.1.1). Österlin et al. (2020) highlight that differing views on wilderness can hinder traditional reindeer herding, as protected areas often prohibit human activities, including those of the Sámi. Consequently, nature conservation measures, such as national parks, restrict Sámi livelihoods (Österlin et al., 2020). This creates a trade-off between the targets on nature conservation (15.1 and 1.4) and cultural heritage (11.4), while other targets in this cluster (6.6, 14.2, and 15.5) support Targets 1.4 and 11.4.

Human activities and economic development in northern Fennoscandia, primarily forestry, mining, and tourism, have significantly impacted the environment (Bäck et al., 2005). Infrastructure like roads, railroads, and buildings has led to habitat loss and affected wildlife. Tourism, in particular, contributes to ecosystem destruction through littering, noise, and the space required for new developments (Heberlein et al., 2002; Bäck et al., 2005). Overall, the impact of economic growth and infrastructure on nature conservation is negative, resulting in trade-offs between targets on economic growth and infrastructure (8.1 and 9.1) and targets on nature conservation (6.6, 14.2, 15.1, and 15.5).

Climate Change Mitigation and Adaptation and the Role of Indigenous Peoples

Climate change is drastically impacting Arctic environments and communities, with warming exceeding the global average in northern latitudes by two times or more (Serreze & Francis, 2006). Indigenous communities, being closely tied to nature, face significant challenges from rising temperatures and their environmental consequences (Toivanen & Cambou, 2021). These include declining sea ice, shifts in ecosystems, extreme weather events, and rising sea levels (Vincent, 2020). Sámi reindeer herders, in particular, experience increased stress in maintaining their culture and traditions due to climate change (Furberg et al., 2011). Thus, mitigating climate change and implementing adaptation measures are crucial for protecting Indigenous economies and cultures, creating synergies between targets 13.3, 1.4, and 11.4. As the warming impacts Arctic environments to such a large degree, reducing greenhouse gas emissions and mitigating climate change also positively impacts biodiversity, eco-systems, and nature conservation in general (Vincent, 2020). This therefore represents another synergy between target 13.3 on climate change and adaptation and nature conservation (Targets 6.6, 14.2, 15.1, and 15.5).

Analysing the impact of wind power projects on Indigenous people in northern Fennoscandia necessitated selecting 14 relevant targets for a synergy and trade-off analysis. The literature review revealed negative effects of wind power on Indigenous communities, particularly traditional reindeer herding, which affects Sámi economies and culture. It also highlighted how these projects undermine Sámi procedural rights and political inclusion, as guidelines for balancing their interests with renewable energy initiatives are lacking. Sámi individuals face discrimination in legal disputes and experience distributional injustice (see Heinämäki et al. 2017).

This creates a trade-off between protecting Sámi culture and access to natural resources and promoting clean energy for fossil-free electricity. Additionally, there are trade-offs between clean energy development and nature conservation due to the environmental impacts of wind turbines. Economic growth and infrastructure improvements also threaten habitat protection and biodiversity. However, synergies exist, as clean energy helps mitigate climate change, which endangers Sámi livelihoods. Furthermore, wind power projects can positively influence economic growth and infrastructure in northern Fennoscandia. Certain interactions are still missing at this phase of the project that this chapter is mainly based on (Arctic Avenue). For instance, the converse effects of nature conservation on economic growth and political inclusion have not been included in this analysis. More focus can also be put on investigating and presenting the specific processes in the study area, such as recent wind power development, in the form of a map. Targets 7.1 and 7.3 have furthermore turned out to not interact with other targets (except for Target 7.2).

In sum, the analysis above shows the complexity of the conflict and how not only climate change affects vulnerable groups (i.e. Indigenous people) but also the impacts of climate change mitigation measures. The chapter's analysis draws attention to the importance of including Indigenous voices and perspectives in decision-making procedures, and raises the question of which system we are allowing to become dominant and what we as a society perceive as worth protecting.

A person from Inari was asked, "Who do you think holds power over issues that concern you, specifically regarding Sámi matters and Sámi reindeer herders? Where are the decisions that affect your life made? Where is the power?" He answered, "In matters concerning Sámi reindeer herding, the Sámi do not hold power; it lies elsewhere. In cultural matters, the Sámi Parliament has authority over culture, language, and other aspects, but when it comes to the reindeer economy and Sámi reindeer herding culture, that power is held by others, not us" (Interview, 2013. Glocal_inari_17male_6.2013, by the author, Finland).

Across the Fennoscandinavian countries, the Sámi people, as Indigenous inhabitants, often find that they have limited influence over how the Sustainable Development Goals (SDGs) are implemented in ways that respect and preserve their culture. This lack of meaningful inclusion risks further marginalisation and cultural harm. Complicating matters is the reality that many Sámi do not live within territories officially recognised as Sámi homelands. As a result, they are not automatically covered by the Free, Prior, and Informed Consent protocols designed to safeguard Indigenous voices.

The Sámi community itself is not monolithic. Regional and cultural distinctions across Sápmi, as well as differing power relations between Sámi groups and state authorities, necessitate nuanced approaches. These complexities must be carefully unpacked when assessing the extent to which Agenda 2030 and its SDGs truly align with human rights principles. This is critical, as the universal claim of the SDGs demands that no one, including Indigenous peoples, be left behind. Yet, without a thorough and inclusive process, the risk remains that these goals may inadvertently perpetuate systemic inequities.

Acknowledgements *I wish to express my gratitude to Hanna Kreplin (then affiliated with the University of Stockholm), for her rigorous and careful work in going through the SDG indicators and carrying out the initial trade-off analysis. I would also like to thank Tuija von der Pütten, Anna Varfolomeeva, and Heidi Sinevaara-Niskanen for their useful comments.

**I wish to thank the Research Council of Finland for funding the Centre of Excellence in Law, Identity and the European Narratives (decision nr. 1353311) without any involvement in the content.

List of SDG Indicators Used in this Chapter

Target	Target Description	Indicator	Indicator Description
1.4	By 2030, ensure that all men and women, in particular the poor and the vulnerable, have equal rights to economic resources, as well as access to basic services, ownership and control over land and other forms of property, inheritance, natural resources, appropriate new technology and financial services, including microfinance	1.4.1	Proportion of population living in households with access to basic services
		1.4.2	Proportion of total adult population with secure tenure rights to land, with legally recognized documentation and who perceive their rights to land as secure, by sex and by type of tenure
6.6	By 2020, protect and restore water-related ecosystems, including mountains, forests, wetlands, rivers, aquifers, and lakes	6.6.1	Change in the extent of water-related ecosystems over time
7.1	By 2030, ensure universal access to affordable, reliable, and modern energy services	7.1.1	Proportion of population with access to electricity
		7.1.2	Proportion of population with primary reliance on clean fuels and technology
7.2	By 2030, increase substantially the share of renewable energy in the global energy mix	7.2.1	Renewable energy share in the total final energy consumption
7.3	By 2030, double the global rate of improvement in energy efficiency	7.3.1	Energy intensity measured in terms of primary energy and GDP
8.1	Sustain per capita economic growth in accordance with national circumstances and, in particular, at least 7 per cent gross domestic product growth per annum in the least developed countries	8.1.1	Annual growth rate of real GDP per capita
9.1	Develop quality, reliable, sustainable, and resilient infrastructure, including regional and trans-border infrastructure, to support economic development and human well-being, with a focus on affordable and equitable access for all	9.1.1	Proportion of the rural population who live within 2 km of an all-season road
		9.1.2	Passenger and freight volumes, by mode of transport
10.2	By 2030, empower and promote the social, economic, and political inclusion of all, irrespective of age, sex, disability, race, ethnicity, origin, religion or economic or other status	10.2.1	Proportion of people living below 50 per cent of median income, by age, sex, and persons with disabilities
10.3	Ensure equal opportunity and reduce inequalities of outcome, including by eliminating discriminatory laws, policies and practices and promoting appropriate legislation, policies and action in this regard	10.3.1	Proportion of the population reporting having personally felt discriminated against or harassed within the previous 12 months on the basis of a ground of discrimination prohibited under international human rights law
11.4	Strengthen efforts to protect and safeguard cultural and natural heritage	11.4.1	Total expenditure (public and private) per capita spent on the preservation, protection, and conservation of all cultural and natural heritage, by type of heritage (cultural, natural, mixed and World Heritage Centre designation), level of government (national, regional and local/municipal), type of expenditure (operating expenditure/investment) and type of private funding (donations in kind, private non-profit sector and sponsorship)

Fig. 4.2 Box with SDG indicators

13.3	Improve education, awareness-raising and human and institutional capacity on climate change mitigation, adaptation, impact reduction and early warning	13.3.1	Number of countries that have integrated mitigation, adaptation, impact reduction and early warning into primary, secondary and tertiary curricula
		13.3.2	Number of countries that have communicated the strengthening of institutional, systemic, and individual capacity-building to implement adaptation, mitigation and technology transfer, and development actions
14.2	By 2020, sustainably manage and protect marine and coastal ecosystems to avoid significant adverse impacts, including by strengthening their resilience, and take action for their restoration in order to achieve healthy and productive oceans	14.2.1	Proportion of national exclusive economic zones managed using ecosystem-based approaches
15.1	By 2020, ensure the conservation, restoration and sustainable use of terrestrial and inland freshwater ecosystems and their services, in particular forests, wetlands, mountains and drylands, in line with obligations under international agreements	15.1.1	Forest area as a proportion of total land area
		15.1.2	Proportion of important sites for terrestrial and freshwater biodiversity that are covered by protected areas, by ecosystem type
15.5	Take urgent and significant action to reduce the degradation of natural habitats, halt the loss of biodiversity and, by 2020, protect and prevent the extinction of threatened species	15.5.1	Red List Index

Fig. 4.2 (continued)

References

Aitken, M. (2010). Wind power and community benefits: Challenges and opportunities. *Energy Policy, 38*(10), 6066–6075. https://doi.org/10.1016/j.enpol.2010.05.062

Aldieri, L., Grafström, J., Sundström, K., & Vinci, C. P. (2020). Wind Power and Job Creation. *Sustainability, 12*(1), 45. https://doi.org/10.3390/su12010045

Bäck, L., Solberg, B., Tömmervik, H., & Wielgolaski, F. E. (2005). Policies and developing plans towards sustainability of mountain birch ecosystems in Scandinavia. In M. M. Caldwell, G. Heldmaier, R. B. Jackson, O. L. Lange, H. A. Mooney, E.-D. Schulze, U. Sommer, F. E. Wielgolaski, P. S. Karlsson, S. Neuvonen, & D. Thannheiser (Eds.), *Plant ecology, herbivory, and human impact in Nordic Mountain birch forests* (pp. 327–340). Springer. https://doi.org/10.1007/3-540-26595-3_24

Bergman, I., Zackrisson, O., & Liedgren, L. (2013). From hunting to herding: Land use, ecosystem processes, and social transformation among Sami AD 800–1500. *Arctic Anthropology, 50*(2), 25–39. https://doi.org/10.3368/aa.50.2.25

Beston, J. A., Diffendorfer, J. E., Loss, S. R., & Johnson, D. H. (2016). Prioritizing avian species for their risk of population-level consequences from wind energy development. *PLoS One, 11*(3). https://doi.org/10.1371/journal.pone.0150813

Bjärstig, T., Nygaard, V., Riseth, J. Å., & Sandström, C. (2020). The institutionalisation of Sami interest in municipal comprehensive planning: A comparison between Norway and Sweden. *The International Indigenous Policy Journal, 11*(2), 1–24. https://doi.org/10.18584/iipj.2020.11.2.10574

Brännlund, I., & Axelsson, P. (2011). Reindeer management during the colonization of Sami lands: A long-term perspective of vulnerability and adaptation strategies. *Global*

Environmental Change, 21(3), 1095–1105. https://doi.org/10.1016/j.gloenvcha.2011.03.005

Broderstad, E. G. (2014). Implementing indigenous self-determination: The case of the Sámi in Norway. In *Restoring indigenous self-determination* (pp. 80–87).

Brown, J. P., Pender, J., Wiser, R., Lantz, E., & Hoen, B. (2012). Ex post analysis of economic impacts from wind power development in U.S. counties. *Energy Economics, 34*(6), 1743–1754. https://doi.org/10.1016/j.eneco.2012.07.010

Cambou, D. (2020). Uncovering injustices in the Green transition: Sámi rights in the development of wind energy in Sweden. *Arctic Review, 11*, 310–333. https://doi.org/10.23865/arctic.v11.2293

Daerga, L., Edin-Liljegren, A., & Sjölander, P. (2008). Quality of life in relation to physical, psychosocial and socioeconomic conditions among reindeer-herding Sami. *International Journal of Circumpolar Health, 67*(1), 10–28. https://doi.org/10.3402/ijch.v67i1.18223

Degai, T. S., & Petrov, A. N. (2021). Rethinking Arctic sustainable development agenda through indigenizing UN sustainable development goals. *International Journal of Sustainable Development & World Ecology, 28*(6), 1–6. https://doi.org/10.1080/13504509.2020.1868608

Ejdemo, T., & Söderholm, P. (2015). Wind power, regional development and benefit-sharing: The case of northern Sweden. *Renewable and Sustainable Energy Reviews, 47*, 476–485. https://doi.org/10.1016/j.rser.2015.03.082

Furberg, M., Evengård, B., & Nilsson, M. (2011). Facing the limit of resilience: Perceptions of climate change among reindeer herding Sami in Sweden. *Global Health Action, 4*(1), 8417. https://doi.org/10.3402/gha.v4i0.8417

Gaultier, S. P., Blomberg, A. S., Ijäs, A., Vasko, V., Vesterinen, E. J., Brommer, J. E., & Lilley, T. M. (2020). Bats and wind farms: The role and importance of the Baltic Sea countries in the European context of power transition and biodiversity conservation. *Environmental Science & Technology, 54*(17), 10385–10398. https://doi.org/10.1021/acs.est.0c00070

Haley, D. (2023). Decolonising regional perspectives for implementing the UN sustainable development goals. In W. Leal Filho, M. A. P. Dinis, S. Moggi, E. Price, & A. Hope (Eds.), *Implementing the UN Sustainable Development Goals – Regional Perspectives* (SDGs in the European region). Springer. https://doi.org/10.1007/978-3-031-17461-2_82

Heberlein, T. A., Fredman, P., & Vuorio, T. (2002). Current tourism patterns in the Swedish Mountain region. *Mountain Research and Development, 22*(2), 142–149. https://doi.org/10.1659/0276-4741(2002)022[0142:CTPITS]2.0.CO;2

Hegre, H., Petrova, K., & von Uexkull, N. (2020). Synergies and trade-offs in reaching the sustainable development goals. *Sustainability., 12*(20), 8729. https://doi.org/10.3390/su12208729

Heikkilä, L., Kuokkanen, R., Lehtola, V-P., Magga, P., Magga, S.-M., Näkkäläjärvi, J., Valkonen, S., & Virtanen, P. (2024). Ethical guidelines for research involving Sámi people in Finland. Giellagas-instituutti. University of Oulu. https://rightsstatements.org/vocab/InC/1.0

Heikkinen, H. (2006). Neo-entrepreneurship as an adaptation model of reindeer herding in Finland. *Nomadic Peoples, 10*(2), 187–208. https://doi.org/10.3167/np.2006.100211

Heinämäki, L., Herrmann, T., & Green, C. (2017). Towards Sámi self-determination over their cultural heritage: The UNESCO world heritage site of Laponia in northern Sweden. In *Indigenous Peoples' Cultural Heritage* (pp. 78–103). https://doi.org/10.1163/9789004342194_005

Helldin, J.-O., Jung, J., Neumann, W., Olsson, M., Skarin, A., Widemo, F., Sverige, Naturvårdsverket, & Vindval (Program). (2012). The impacts of wind power on terrestrial mammals*: A synthesis*. Naturvårdsverket.

Henriksen, J. B. (2008). The continuous process of recognition and implementation of the Sami people's right to self-determination. *Cambridge Review of International Affairs*. https://doi.org/10.1080/09557570701828402

Jaakkola, J. J. K., Juntunen, S., & Näkkäläjärvi, K. (2018). The holistic effects of climate change on the culture, well-being, and health of the Saami, the only indigenous people in the European Union. *Current Environmental Health Reports, 5*(4), 401–417. https://doi.org/10.1007/s40572-018-0211-2

Kaiser, N., Sjölander, P., Liljegren, A. E., Jacobsson, L., & Renberg, E. S. (2010). Depression and anxiety in the reindeer-herding Sami population of Sweden. *International Journal of Circumpolar Health, 69*(4), 383–393. https://doi.org/10.3402/ijch.v69i4.17674

Keskitalo, C. (2007). International region-building: Development of the Arctic as an international region. *Cooperation and Conflict, 42*(2), 187–205. https://doi.org/10.1177/0010836707076689

Kreplin, H. N., Santos Ferreira, C. S., Destouni, G., Keesstra, S. D., Salvati, L., & Kalantari, Z. (2021). Arctic wetland system dynamics under climate warming. *WIREs Water, 8*, e1526. https://doi.org/10.1002/wat2.1526

Larsen, R. K., & Raitio, K. (2019). Implementing the state duty to consult in land and resource decisions: Perspectives from Sami communities and Swedish state officials. *Arctic review, 10*, 4–23. https://doi.org/10.23865/arctic.v10.1323

Larsen, R. K., Raitio, K., Stinnerbom, M., & Wik-Karlsson, J. (2017). Sami-state collaboration in the governance of cumulative effects assessment: A critical action research approach. *Environmental Impact Assessment Review, 64*, 67–76. https://doi.org/10.1016/j.eiar.2017.03.003

Lawrence, R. (2014). Internal colonisation and indigenous resource sovereignty: Wind power developments on traditional Saami lands. *Environment and Planning D, 32*(6), 1036–1053. https://doi.org/10.1068/d9012

Lawrence, R., & Åhrén, M. (2016). Mining as colonisation: The need for restorative justice and restitution of traditional Sami lands. In *Nature, Temporality and Environmental Management*. Routledge.

Leu, T. C., Eriksson, M., & Müller, D. K. (2018). More than just a job: Exploring the meanings of tourism work among indigenous Sámi tourist entrepreneurs. *Journal of Sustainable Tourism, 26*(8), 1468–1482. https://doi.org/10.1080/09669582.2018.1466894

Liljenfeldt, J. (2013). *Wind power development as a means to local economic development* (pp. 124–141). Institutionen för geografi och ekonomisk historia, Umeå universitet. http://urn.kb.se/resolve?urn=urn:nbn:se:umu:diva-85146

Liljenfeldt, J., & Pettersson, Ö. (2017). Distributional justice in Swedish wind power development—An odds ratio analysis of windmill localization and local residents' socioeconomic characteristics. *Energy Policy, 105*, 648–657. https://doi.org/10.1016/j.enpol.2017.03.007

May, R., Jackson, C. R., Middel, H., Stokke, B. G., & Verones, F. (2021). Life-cycle impacts of wind energy development on bird diversity in Norway. *Environmental Impact Assessment Review, 90*, 106635. https://doi.org/10.1016/j.eiar.2021.106635

Müller-Wille, L., Heinrich, D., Lehtola, V.-P., Aikio, P., Konstantinov, Y., & Vladimirova, V. (2006). Dynamics in human-reindeer relations: Reflections on prehistoric, historic and contemporary practices in northernmost Europe. In B. C. Forbes, M. Bölter, L. Müller-Wille, J. Hukkinen, F. Müller, N. Gunslay, & Y. Konstantinov (Eds.), *Reindeer Management in Northernmost Europe: Linking practical and scientific knowledge in social-ecological systems* (pp. 27–45). Springer. https://doi.org/10.1007/3-540-31392-3_3

Munday, M., Bristow, G., & Cowell, R. (2011). Wind farms in rural areas: How far do community benefits from wind farms represent a local economic development opportunity? *Journal of Rural Studies, 27*(1), 1–12. https://doi.org/10.1016/j.jrurstud.2010.08.003

Natcher, D. C., & Koivurova, T. (Eds.). (2021). *Renewable economies in the Arctic* (1st ed.). Routledge. https://doi.org/10.4324/9781003172406

Nilsson, A. E., & Larsen, J. N. (2020). Making regional sense of global sustainable development indicators for the Arctic. *Sustainability, 12*(3), 1027. https://doi.org/10.3390/su12031027

Normann, S. (2021). Green colonialism in the Nordic context: Exploring southern Saami representations of wind energy development. *Journal of Community Psychology, 49*(1), 77–94. https://doi.org/10.1002/jcop.22422

Nuttall, M. (2018). Arctic environments and peoples. In *The international encyclopedia of anthropology* (pp. 1–7). American Cancer Society. https://doi.org/10.1002/9781118924396.wbiea1480

Nuttall, M. (2021). Arctic ecology, indigenous peoples and environmental governance. In D. N. Thomas (Ed.), *Arctic ecology*. https://doi.org/10.1002/9781118846582.ch15

Nysten-Haarala, S., Joona, T., & Hovila, I. (2021). Wind energy projects and reindeer herders' rights in Finnish Lapland: A legal framework. *Elementa, 9*(1), 00037.

Österlin, C., & Raitio, K. (2020). Fragmented landscapes and Planscapes—The double pressure of increasing natural resource exploitation on indigenous Sámi lands in northern Sweden. *Resources, 9*(9), 104. https://doi.org/10.3390/resources9090104

Österlin, C., Schlyter, P., & Stjernquist, I. (2020). Different worldviews as impediments to integrated nature and cultural heritage conservation management: Experiences from protected areas in northern Sweden. *Sustainability, 12*(9), 3533. https://doi.org/10.3390/su12093533

Pradhan, P., Costa, L., Rybski, D., Lucht, W., & Kropp, J. P. (2017). A systematic study of sustainable development goal (SDG) interactions. *Earth's Future, 5*(11), 1169–1179. https://doi.org/10.1002/2017EF000632

Raitio, K., Allard, C., & Lawrence, R. (2020). Mineral extraction in Swedish Sápmi: The regulatory gap between Sami rights and Sweden's mining permitting practices. *Land Use Policy, 99*. https://doi.org/10.1016/j.landusepol.2020.105001

Ren, G., Liu, J., Wan, J., Li, F., Guo, Y., & Yu, D. (2018). The analysis of turbulence intensity. Based on wind speed data in onshore wind farms. *Renewable Energy, 123*, 756–766. https://doi.org/10.1016/j.renene.2018.02.080

Salmi, A.-K., & Heino, M. T. (2019). Tangled worlds: The Swedish, the Sámi, and the reindeer. *International Journal of Historical Archaeology, 23*(1), 260–282. https://doi.org/10.1007/s10761-018-0465-2

Sara, M. N. (2009). Siida and traditional Sámi reindeer herding knowledge. *Northern Review, 30*, 153–178.

Selle, P., & Strømsnes, K. (2010). Sámi citizenship: Marginalisation or integration? *Acta Borealia, 27*(1), 66–90. http://www.tandfonline.com/doi/abs/10.1080/08003831.2010.486926

Serreze, M. C., & Francis, J. A. (2006). The Arctic amplification debate. *Climatic Change, 76*(3–4), 241–264.
Simpson, A. (2020). The sovereignty of critique. *South Atlantic Quarterly, 119*(4), 685–699. https://doi.org/10.1215/00382876-8663591
Skarin, A., & Åhman, B. (2014). Do human activity and infrastructure disturb domesticated reindeer? The need for the reindeer's perspective. *Polar Biology, 37*(7), 1041–1054. https://doi.org/10.1007/s00300-014-1499-5
Skarin, A., Nellemann, C., Rönnegård, L., Sandström, P., & Lundqvist, H. (2015). Wind farm construction impacts reindeer migration and movement corridors. *Landscape Ecology, 30*(8), 1527–1540. https://doi.org/10.1007/s10980-015-0210-8
Skarin, A., Sandström, P., & Alam, M. (2018). Out of sight of wind turbines—Reindeer response to wind farms in operation. *Ecology and Evolution, 8*(19), 9906–9919. https://doi.org/10.1002/ece3.4476
Sköld, P., Baer, K. C., Scheepstra, A., Latola, K., & Biebow, N. (2018). The SDGs and the Arctic: The need for polar indicators. Paper presented at Arctic observing summit 2018, Davos, Switzerland.
Sovacool, B. K. (2013). The avian benefits of wind energy: A 2009 update. *Renewable Energy, 49*, 19–24. https://doi.org/10.1016/j.renene.2012.01.074
Stepien, A., Koivurova, T., Gremsperger, A., & Niemi, H. (2014). Arctic indigenous peoples and the challenge of climate change. In E. Tedsen, S. Cavalieri, & R. A. Kraemer (Eds.), *Arctic marine governance: Opportunities for transatlantic cooperation* (pp. 71–99). Springer. https://doi.org/10.1007/978-3-642-38595-7_4
TENK. (2019). Guidelines for ethical review in human sciences. Finnish National Board on research integrity 3/2019 (2nd revised ed.). https://tenk.fi/sites/default/files/2021-01/Ethical_review_in_human_sciences_2020.pdf
The Law on Sámi Parliament. (974/1995). Amendments up to 1026/2003 included, [laki saamelaiskäräjistä].
Thomas, D. ed, 2020. *Arctic Ecology*. Oxford: Wiley Blackfell.
Toivanen, R. (2015). Obstacles and successes. In H. F. Marten, M. Rießler, J. Saarikivi, & R. Toivanen (Eds.), *Cultural and linguistic minorities in The Russian Federation and the European Union: Comparative studies on equality and diversity* (pp. 83–105). Springer International Publishing. https://doi.org/10.1007/978-3-319-10455-3_4
Toivanen, R., & Cambou, D. (2021). Human rights. In C. P. Krieg & R. Toivanen (Eds.), *Situating sustainability: A handbook of concepts and contexts* (pp. 51–62). Helsinki University Press. https://hup.fi/site/chapters/e/10.33134/HUP-14-4/
Toivanen, R., & Fabritius, N. (2020). Arctic youth transcending notions of 'culture' and 'nature': Emancipative discourses of place for cultural sustainability. *Current Opinion in Environmental Sustainability, 43*, 58–64.
Toivanen, R. J. (2019). European Fantasy of the Arctic region and the rise of indigenous Sámi voices in the global arena. In N. Sellheim, Y. Zaika, & I. Kelman (Eds.), *Arctic Triumph. Springer Polar Sciences.* Springer. https://doi.org/10.1007/978-3-030-05523-3_3
Tuck, E., & McKenzie, M. (2014). *Place in research: Theory, methodology, and methods.* Routledge.
Tuhiwai Smith, L. (2012). Decolonizing Methodologies: Research and Indigenous Peoples. Zed Books.
UN (2015). Transforming our world: The 2030 agenda for sustainable development. General Assembly A/RES/70/1.

Virtanen, P. K., & Seurujärvi-Kari, I. (2019). Introduction: Theorizing Indigenous Knowledge(s). *Dutkansearvvi dieđalaš áigečála, 3*(2), 1–19. http://dutkansearvi.fi/wp-content/uploads/2020/02/Dutkansearvi-Vol3Issue2-Virtanen-SeurujarviKari.pdf

Vincent, W. F. (2020). Arctic climate change: Local impacts, global consequences, and policy implications. In K. S. Coates & C. Holroyd (Eds.), *The Palgrave handbook of Arctic policy and politics* (pp. 507–526). Springer International Publishing. https://doi.org/10.1007/978-3-030-20557-7_31

Wang, S., & Wang, S. (2015). Impacts of wind energy on environment: A review. *Renewable and Sustainable Energy Reviews, 49*, 437–443. https://doi.org/10.1016/j.rser.2015.04.137

Wang, S., Wang, S., & Smith, P. (2015). Ecological impacts of wind farms on birds: Questions, hypotheses, and research needs. *Renewable and Sustainable Energy Reviews, 44*, 599–607. https://doi.org/10.1016/j.rser.2015.01.031

Weitz, N., Carlsen, H., Nilsson, M., & Skånberg, K. (2018). Towards systemic and contextual priority setting for implementing the 2030 agenda. *Sustainability Science, 13*(2), 531–548. https://doi.org/10.1007/s11625-017-0470-0

Williams, S. M. (2003). Tradition and change in the sub-Arctic: Sámi reindeer herding in the modern era. *Scandinavian Studies, 75*(2), 229–256.

Yap, M. L.-M., & Watene, K. (2019). The sustainable development goals (SDGs) and indigenous peoples: Another missed opportunity? *Journal of Human Development and Capabilities, 20*(4), 451–467. https://doi.org/10.1080/19452829.2019.1574725

Zhou, L., Tian, Y., Baidya Roy, S., Thorncroft, C., Bosart, L. F., & Hu, Y. (2012). Impacts of wind farms on land surface temperature. *Nature Climate Change, 2*(7), 539–543. https://doi.org/10.1038/nclimate1505

CHAPTER 5

Clean Energy Investments and Indigenous Peoples' Rights: Sustainability Issues in Wind Farm Projects

Tuija von der Pütten

Introduction

Although investments in renewable energy play a crucial role in achieving sustainable development goals (SDGs) and global net-zero emissions in accordance with the Paris Agreement, there is growing concern about the adverse human rights impacts linked to renewable energy projects, particularly concerning Indigenous peoples (Loyer et al., 2021; Fjellheim, 2023a; Agrawal et al., 2023). Without effective management of their human rights impacts, renewable energy companies are at risk of causing significant harm to Indigenous communities, which could lead to increased legal risks, damage the sector's credibility, and diminish its positive contributions to sustainability and efforts at addressing climate change. Renewable energy projects have been linked to hundreds of allegations of human rights abuses across the globe (the Business and Human Rights Resource Centre, BHRRC, 2021a), and as a result calls for a just transition have emerged (Cambou, 2020; Agrawal et al., 2023). Many Indigenous communities are demanding their legal right to self-determination and Free, Prior and Informed Consent (FPIC) as well as the need to prioritise Indigenous knowledge in the green transition (Young, 2019; Schilling-Vacaflor & Flemmer, 2020; O'Neill et al., 2021). FPIC, an idea emerging within Indigenous discourse as a mechanism to safeguard Indigenous rights, is currently recognised in various international legal instruments. The concept refers

T. von der Pütten (✉)
Faculty of Law & Helsinki Institute for Sustainability Science (HELSUS), University of Helsinki, Helsinki, Finland
e-mail: tuija.vonderputten@helsinki.fi

R. Toivanen et al. (eds.), *Decolonizing the Sustainable Development Goals*,
https://doi.org/10.1007/978-3-032-14263-4_5

to the right of Indigenous peoples to give or withhold their consent for actions that would affect their lands or territories. Regrettably, although the mode of business operation has been changing from that of extractive industry, or 'brown' fossil fuel industry, to more sustainable modes of energy production, the power relations between companies, local authorities and Indigenous people remains the same. The 'historic inequality in legal rights and continuing inequities in socio-economy, health and wellbeing' still puts Indigenous people at greater risk across all aspects of energy justice (O'Neill et al., 2021). Consequently, the expansion of the renewable energy industry has at times led to land acquisition, creating 'green sacrifice zones' (Smith & Scott, 2022, p. 237; Zografos, 2020) and 'green colonialism' (Kårtveit, 2021; Fjellheim, 2023b), which calls the sustainability of renewable energy projects into question. Although Indigenous communities may overall support the efforts at combating climate change, they resist the idea that they must sacrifice their lands to serve the needs of global renewable energy investors and developers (O'Neill et al., 2021).

This chapter explores how renewable energy companies and their investors are implicated in projects that violate Indigenous peoples' rights. Although clean investments are crucial for ensuring access to affordable and clean energy (SDG 7) and for taking action against climate change (SDG 13), such activities may be in conflict with Indigenous peoples' participatory rights (SDG 16). Using a case study approach, this chapter examines three wind farm projects on Indigenous peoples' lands: Fosen in Norway, Lake Turkana in Kenya and Gunaa Sicarú in Mexico. The case studies were selected because all are large-scale green investments involving complicated juridic processes with Indigenous peoples' communities. The chapter demonstrates the extent to which the companies involved in the projects are violating Indigenous peoples' right to FPIC and explores the legal implications of the violations for investors. More particularly, I look at how court cases or other complaints by Indigenous people address investor involvement in the violations. While the obligation to protect and to fulfil human rights rests with states, businesses (and their investors) are responsible for not harming people or causing adverse human rights impacts (see UN Guiding Principles on Business and Human Rights). Similarly, although the responsibility to guarantee FPIC and consultation rests with nation-states, corporations often wield significant influence in determining if and how they are implemented and whether nation-states adhere to this right (Ewell, 2022). The existing critical literature suggests that the voluntary duty of companies to respect human rights creates severe difficulties for protecting those rights (Deva, 2012; Young, 2019). Moreover, states and companies may disregard the rights when they conflict with economic interests. This is also evident in clean energy investment projects on Indigenous peoples' lands, which calls for more ambitious efforts to make the green transition truly sustainable.

Clean Energy Investments, Indigenous Peoples and the Right to FPIC

The urgent need to transition to a carbon-neutral economy necessitates significant public and private investments in renewable energy, particularly in wind and solar energy. Since the adoption of the Paris Agreement, international investments in renewable energy have nearly tripled (UNCTAD, 2023). The International Renewable Energy Agency (IRENA) estimates that more than USD 5 trillion per year is required to achieve the 1.5 °C climate target by 2050 (IRENA, 2023). International and domestic investors are already financing renewable energy installations in many parts of the world, of which state-owned enterprises and development banks form a large part. Additionally, foreign direct investments by multinational companies play a significant role in renewable energy investments (Knutsson & Flores, 2022).

Yet, while the world is slowly heading towards a green transition, there have been increasing allegations and lawsuits against wind and solar companies for contributing to or causing the displacement of local peoples or for their involvement in projects with adverse human rights impacts. Many of the impacts affect Indigenous peoples and local communities. The Business and Human Rights Resource Centre (BHRRC) has documented more than 200 allegations of adverse human rights impacts within the renewable energy industry between 2010 and 2020. Notably, 44% of those allegations were associated with the wind and solar sectors (BHRRC, 2021a). As of May 2024, the BHRRC's Just Transition Litigation Tracking Tool had documented 60 legal cases launched by Indigenous peoples and local communities against companies or states linked to transition mineral mining or the renewable energy sector (BHRRC, 2024).

Wind and solar energy companies often progress large-scale developments covering vast areas of land, which require up to 10 times more land per unit of produced power compared to fossil fuels and threaten the culture, existence and livelihoods of the communities living on that land (Williams, 2023). Tensions between clean energy developments and Indigenous peoples arise predominantly due to the (i) land-intensive nature of renewable energy projects, (ii) insufficient human rights due diligence by both governments and business enterprises in the planning and execution of extensive projects, and (iii) unequal power structures between companies, Indigenous people and governments (OHCHR, 2017; Eisenberg, 2020). The situation is complicated in areas inhabited by Indigenous peoples, where legal rights to the land are superimposed on pre-existing ownership structures. Despite representing only 6% of the global population, Indigenous peoples hold management or tenure rights to over a quarter of the world's land across 90 countries (Williams, 2023). Unfortunately, such rights are seldom officially acknowledged, underscoring the risk of infringing on Indigenous peoples' collective rights to traditional land and self-determination. Especially Indigenous people practising a nomadic lifestyle are in a problematic position when states sell land to private businesses and the local Indigenous people lose their free access to use of that land.

Although Indigenous rights to practise their culture and Indigenous land rights, as well as their right to self-determination and right to FPIC, are protected by international treaties, declarations and principles, such as the ILO Convention No. 169, International Covenant on Civil and Political Rights (ICCPR), UN Declaration on the Rights of Indigenous Peoples (UNDRIP) and UN Guiding Principles on Business and Human Rights (UNGPs), these rights are frequently disregarded at the national level, whether by states or private entities.

The right to FPIC has become an important norm in international law (Young, 2019), and it refers to the need to obtain consent from Indigenous peoples for any activities undertaken on their land or in their territory. It covers issues pertaining to the right to self-determination and consultation in decision-making processes of relevance to them (Åhrén, 2016). The private sector's response to this obligation is referred to as 'social license to operate' (Stuart et al., 2023). Enforcing the right, however, is challenging due to imbalanced power structures between companies, governments and Indigenous communities. Indigenous peoples and local communities frequently lack the political leverage to enforce their rights without governmental support (Young, 2016; Roesch, 2016). States and multinational corporations' interests are also generally misaligned with the legal recognition of FPIC (Young, 2016). Additionally, application of the right presents challenges because definitions of indigeneity, historical conflicts over land rights and colonial legacies vary across regions (Ewell, 2022).

Although international legal sources primarily focus on the role of governments in obtaining FPIC, business enterprises play an important role in determining how and whether the right is upheld. When international law, EU law, declarations, principles and guidelines, as well as case law, are viewed holistically, it is clear that the rights of Indigenous peoples have implications for both states and companies (Lehr & Smith, 2010). For example, according to UNDRIP, states may delegate the responsibility for obtaining FPIC to companies. The EU Sustainability Reporting Standard (EU/2023/2772) stipulates that companies need to take FPIC into consideration in their materiality assessment.[1] The EU Sustainable Finance Disclosure Regulation (EU/2019/2088) also imposes requirements on investors concerning Environmental, Social and Governance (ESG) risks and any adverse impacts on their investments. Companies are under increasing investor-driven pressure to proactively implement measures to mitigate ESG impacts, including those violating the human rights of local communities. A failure to comply with the terms may constitute a breach of contract and may even lead to termination of the project finance agreement (Dolton-Zborowski & Szoke-Burke, 2022). Further, various corporate social responsibility networks, such as the UN Global Compact, and

[1] Appendix A, AR. 16, Council Directive 2023/2772.

industry associations have adopted FPIC guidelines. These initiatives frequently reference the United Nations Guiding Principles on Business and Human Rights, a nonbinding framework that outlines corporate responsibilities to uphold international human rights law.

Additionally, investors and developers are increasingly integrating FPIC into their own human rights due diligence standards. Many regional development banks, the World Bank and the International Finance Corporation (IFC) expressly mandate that companies acquire FPIC. The Equator Principles, the risk management framework for financial institutions, as well as the Equitable Origin EO100, an industry certification scheme focusing on renewable energy projects, require that clients respect Indigenous peoples' rights, including Free, Prior and Informed Consent (Equator Principles, 2020; Equitable Origin, 2024). Also, the world's largest asset manager, BlackRock, encourages companies to obtain FPIC (BlackRock, 2024).

Common violations include land acquisition without FPIC, displacement without compensation, the loss of culture, traditions or livelihood, and violence against human rights defenders (Tzay, 2023). These violations are exacerbated when foreign investors are involved (Special Rapporteur, 2016). Renewable energy projects have also faced significant backlash when there has been insufficient engagement with Indigenous peoples and when FPIC has not been genuinely obtained. For example, failures to secure community approval may give rise to prolonged social unrest and conflict, as seen in the Niger Delta, where the neglect of Indigenous concerns by a foreign oil company sparked decades of conflict. Financiers may also cancel their funding, as happened, for instance, with a geothermal project in Kenya (see Danwatch, 2022).

Legal risks arising from the adverse human rights impacts caused by renewable energy projects include potential exposure to community litigation. Such litigation can have financial, operational and reputational consequences, including project delays, modifications and fines. Companies have already faced lawsuits from communities in host country courts involving breaches of laws relating to land rights, consultation and FPIC (Rambouts, 2014). Additionally, companies and investors may face nonjudicial complaints through the grievance mechanism established under the Organisation for Economic Co-operation and Development (OECD) Guidelines as well as various development finance institutions, certification schemes and other multi-stakeholder or sustainability-focused initiatives.

Renewable energy companies are expected to operate sustainably, including through conducting impact assessments, but as the empirical reality demonstrates, the vagueness and malleability of human rights law as well as a lack of legal rules pertaining to business and human rights complicates the proper implementation of Indigenous peoples' rights. Renewable energy projects developed by corporations and investors seeking profit can easily avoid human rights and environmental responsibilities due to complicated corporate structures and/or due to the voluntary nature of human rights standards (Muchlinski, 2014), despite the increasing legality of human rights due diligence (Deva,

2023). Foreign investors especially may dodge liability for human rights violations or environmental harm caused to Indigenous peoples by their operations (Special Rapporteur, 2016; Perrone, 2022). While the duty to protect Indigenous rights primarily lies with the state, companies and investors also have duties and responsibilities to respect their rights. Although it is challenging to completely resolve conflicts between companies and Indigenous peoples, there is a growing acknowledgement of the crucial role that the private sector plays in upholding the rights of Indigenous peoples.

Fosen Wind Farm in Sámi Grazing District, Norway

The Fosen case concerns the construction of the Storheia and Roan wind farms on the Fosen peninsula, where Sámi *siidas* practise reindeer husbandry. The Fosen case began in 2010 when the Norwegian Water Resources and Energy Directorate issued licences and an expropriation decision in favour of the windfarms to Fosen Vind DA, a joint venture between the Norwegian state-owned enterprise Statkraft, the European investor consortium Nordic Wind Power and TrønderEnergi. The plan was to build 151 windmills that would provide renewable energy, coinciding with SDG 7 and SDG 13 (Statkraft, 2024). The Fosen Sámi communities opposed the project from the start, fearing damage to their ancestral lands and pastures. They attempted to stop the construction, first by filing a complaint with the Ministry of Petroleum and Energy (OED) and later by lodging a case at the District Court, then disputing the decision at the court of appeal and bringing the case to the Supreme Court of Norway. The ministry rejected all complaints and the project was allowed to proceed, ultimately being completed in 2019 and 2020.

The Sámi argued at all court levels that the project violated their right to enjoy their culture in accordance with Article 27 of the ICCPR,[2] Article 1 of Protocol 1 of the ECHR[3] and Article 5 of the ICERD.[4] The Sámi Parliament argued that neither they nor the Fosen Sámi community had provided their free, prior and informed consent for the projects (Saami Council, 2023). The district court found that the Sámi had lost their grazing lands as a result and ordered Fosen Vind to compensate them. Both the Sámi and Fosen Vind appealed the decision. While Fosen Vind argued that the compensation was too high, the Sámi community maintained that the wind farm violated their rights as established under international law. The court of appeal increased the amount of compensation due to the loss of winter-feeding land. Both parties appealed the decision to the Supreme Court of Norway. In 2021, the Supreme Court of Norway ruled that construction of the wind farm violated the

[2] International Covenant on Civil and Political Rights (adopted 19 December 1966, entered into force 23 March 1976) 999 UNTS 171.

[3] European Convention for the Protection of Human Rights and Fundamental Freedoms, CETS 005, Council of Europe, 1950.

[4] International Convention on the Elimination of All Forms of Racial Discrimination (adopted 21 December 1965, entered into force 4 January 1969), 660 UNTS 195.

reindeer herders' right to enjoy their culture under Article 27 of the ICCPR, and the issued licence was therefore rendered invalid (Supreme Court of Norway, 2021, para. 153). Although neither the Sámi nor the court referred to FPIC specifically, the court acknowledged the significance of consultation as established under international and Norwegian law (Supreme Court of Norway, paras. 120–122.)

In the proceedings of the court of appeal and the Supreme Court of Norway, the Sámi brought forth claims against Fosen Vind that related to the environmental impact assessment (EIA) and lack of consultation with the Sámi. Fosen Vind had been issued the final licence based on an EIA that the company had commissioned. The consultants concluded that the wind park would not have a negative impact on reindeer husbandry, completely ignoring Indigenous knowledge and the concerns of local reindeer herders (Fjellheim, 2023a; Society for Threatened Peoples, 2018). Since the OED had granted the licence, it merely affirmed that the wind park would not breach Article 27 of the ICCPR, which was a miscalculated risk regarding the scope of the article (Ravna, 2023, p. 158). However, the lawyer representing the Fosen reindeer herders contended that the EIA commissioned by Fosen Vind was 'superficial', ignored 'the knowledges and concerns of reindeer herders', and failed 'to assess the cumulative impacts on reindeer herding landscapes from other encroachments' (Fjellheim, 2023a, p. 149). Although the OED was responsible for approving the licence for the project, Fosen Vind, as the developer, was responsible for pushing the project forward, despite Sámi opposition, for commissioning the EIA, and for producing 'doubt about all knowledge which threatened their commercial interests' (Fjellheim, 2023a, p. 160).

The EIA process has been heavily criticised for being an industry-led process, for adopting a 'business-friendly' interpretation of the existing legislation, and for being based on environmental sciences, which in this instance did not take into consideration traditional Sámi knowledge (Fjellheim, 2023a). According to the Sámi scholar Eva Fjellheim (2023a), Fosen Vind downplayed the observed impacts on reindeer herders and leaned on information generated by consultants, who adjusted the EIA to better align with the preferences of the licensing authorities. Fjellheim found 'epistemic miscommunication' between Sámi and consultants conducting the EIA as well as 'state and corporate domination' over reindeer herders' traditional knowledge (Fjellheim, 2023b, p. 27).

The Norwegian government, as Statkraft's owner, had an interest in a favourable judgement for Fosen Vind. Simultaneously, the Norwegian government bears the legal responsibility to uphold and protect the human rights of the Sámi. The balance appeared to heavily favour economic interests stemming from the wind farm's operation. In this context, the conflicts between reindeer herders, the state and companies underscore disputes regarding the type of knowledge prioritised in certain instances and whose perspectives should shape the assessment of impacts (Cambou, 2024). The disputes also involve, as Fjellheim notes, clashes in ontologies, or worldviews, as different parties consider what is at stake when large infrastructure projects disrupt Sámi landscapes (Fjellheim, 2023a). States are accountable for ensuring that consultations take place, but in practice companies tend to disrespect such measures or even commit acts of violence and coercion to achieve what they want (Dunlap, 2018; Fjellheim, 2023a). On a similar note, Dunlap views consultation and FPIC procedures as a type of inclusive control that 'enforces colonial law and safeguards corporate investments' (Dunlap, 2018, p. 105).

In addition to the court cases, the South Sámi took actions against Fosen Vind and its financiers through non-juridical complaint mechanisms. At the same time as the case was being tried in the Supreme Court of Norway, the Sámi, through the NGO Society for Threatened Peoples (STP), put pressure on the investor, Nordic Wind Power DA, and on Swiss financing companies to cease the project, withdraw their investments and ensure the Sámi people's right to FPIC (Renewables Now, 2020). Nordic Wind Power DA is a consortium of European investors initially established by Credit Suisse Energy Infrastructure Partners AG, which had a 40% share in Fosen Vind. BKW Energie AG, headquartered in Bern, is also part of the consortium, holding an indirect 11.2% stake in the project. To address these concerns, in December 2018 the STP organised a visit to Switzerland, enabling the South Sámi to engage in direct discussions with BKW and Credit Suisse investors. Subsequently, the STP filed a complaint against BKW with the Swiss National Contact Point (NCP) for the OECD Guidelines for Multinational Enterprises (see *Society for Threatened Peoples Switzerland* vs. *BKW Group*, 2020). The Swiss NCP accepted the specific reasons for the complaint and organised mediation between the parties. As a result, BKW agreed to revise its code of conduct and to incorporate respect for human rights and the principle of FPIC into its internal guidelines (OECD Watch, 2020). BKW also agreed to develop its human rights due diligence and implement FPIC during the lifecycle of projects, as well as to make its complaint mechanism accessible to all groups at the project level (OECD Watch, 2020; Society for Threatened Peoples, 2023).

The Fosen case has direct implications for Fosen Vind as well as for future clean energy concessions in Norway. Statnett and Fosen Vind were ordered to pay damages and legal fees to the Sámi. The government and the developer were also responsible for finding a way to redress the damages. The court maintained that the windmills' alignment with Norway's environmental objectives for cleaner energy was not a pertinent factor in assessing whether their installation plans infringed upon Indigenous rights. Therefore, any establishment of a wind turbine park that significantly hinders or restricts the Sámi communities' engagement in reindeer husbandry constitutes a violation of their rights. Accordingly, following the Supreme Court of Norway's rationale, concessions for windmills can only be granted when the installation does not impede Indigenous people from practising their cultural rights in accordance with Article 27 of the ICCPR. The court determined that no substantial conflict existed between environmental priorities and Indigenous rights in this particular instance, as alternative locations in Norway were available for the installation of wind turbines without encroaching upon the rights of the Sámi people (Supreme Court of Norway, 2021, para 143). After the court ruling, the Norwegian government apologised to the Sámi (Rasmussen, 2023; Government of Norway, 2023). The government, Statkraft and the Sámi communities have subsequently been involved in a mediation process, and a partial solution has been found. Statkraft has agreed to pay 7 million Norwegian crowns to the *siidas* to the south of Fosen each year, for 25 years (Rasmussen, 2023), and the government has agreed to allocate alternative grazing lands to them (Christiansen, 2024). Fosen Vind's reputation has been severely tarnished, and it has been accused of engaging in green colonialism (Normann, 2020; Fjellheim, 2023b).

Lake Turkana Wind Farm in Marsabit County, Kenya

The second case study centres on the Lake Turkana wind farm, a 310 MW wind energy project with 365 wind turbines, located in Marsabit County, Kenya, and built between 2014 and 2016. The location of the wind park is one of the most remote and poorest regions in Kenya. Most of the population are pastoralists of different ethnic backgrounds, mainly Boran, Gabra, Rendille, Samburu, Somali and Turkana groups, who practise livestock rearing (Lomax et al., 2023). The project is based in a region widely acknowledged for violence and conflict between communities, violence particularly between Turkana and Samburu groups (Lomax et al., 2023). The wind farm was financed by the European Investment Bank, African Development Bank, Netherlands Development Finance Company and other smaller investors. It is currently owned by six shareholders, including Anergi Turkana Investments Limited, Milele Energy Ltd., BlackRock and the Danish Climate Fund (Finnfund, 2023; Lake Turkana Wind Project, 2024). Five major contractors were responsible for the project: Vestas, Siemens, SECO, Civicon and RXPE. The wind park is one of the largest private investments in Kenya and the largest wind farm in Africa, consisting of 365 wind turbines.

The government of Kenya leased the land (including the transfer of rights to the trust land) to the Lake Turkana Wind Power Ltd. (LTWP) consortium to develop the windmill park. This led to heightened tensions in the region, particularly due to conflicting assertions regarding the rangeland in Sarima. In 2014, local communities and pastoralists in the Lake Turkana region filed a case with the Environment and Land Court at Meru against LTWP. The claim alleged that the company had not respected their rights when leasing the land, including the right to consultation and FPIC. The claimants sought nullification of the lease and termination of the project. They alleged that the territory where the wind park was being constructed was part of their ancestral lands and that their displacement without adequate compensation or consent threatened their cultural heritage, traditional livelihood and survival (Environment and Land Court, 2021, para 3). Further, they claimed that nomadism was integral to their lifestyle and that the privatisation of the land resulted in the loss of cultural sites and restricted access for livestock and camels to the waters of Lake Turkana. The Lake Turkana Indigenous communities referenced the Banjul Charter, ICCPR, ICESCR, ILO Convention No. 169 and the IFC Performance Standards, asserting that, as Indigenous inhabitants of Marsabit County, they are entitled to protection under these international agreements (Environment and Land Court, 2021, para 5).

The LTWP denied their claims, arguing that some of the claimants were not residents of Marsabit County and that the land did not culturally belong to such communities (Environment and Land Court, 2021, para. 16). Moreover, the company argued that the claimants were not seeking compensation because they recognised that their use of the land would not be affected. One of the LTWP respondents also noted that the project is important for promoting sustainable development with great socio-economic benefits to the local communities (Environment and Land Court, 2021, para 23).

In its 2021 decision, the court acknowledged the claimants' argument that the approval process had violated their rights in accordance with international law and the IFC Performance Standards regarding FPIC (Environment and Land Court, 2021, paras. 4, 5 and 110). The court noted that the consortium did not notify the relevant divisional board under Kenya's Trust Land Act to set the land aside for development, and therefore no proper public consultation had taken place. If the consortium had notified the divisional board, the action would have been followed by a proper public consultation. Although the court did not identify a breach of the cultural, environmental or social rights safeguarded by the Constitution of Kenya for the impacted communities (Environment and Land Court, 2021, para. 130), and neither did it address the indigeneity of the plaintiffs or the difference between Indigenous and tribal peoples, it did recognise the entitlement of the tribal groups affected by the wind park to bring a lawsuit based on their ancestral rights and interests in the land. The court also highlighted that the LTWP consortium, as the investor in the project, 'should have undertaken careful and foolproof due diligence' (Environment and Land Court, 2021, para. 157). Nevertheless, the court ultimately ruled that the title deeds for the land on which the windmills were constructed were unlawful, but it did not order the wind farm to discontinue operations by acknowledging that the project had already been completed and was supplying a substantial amount of electricity. Instead, it ordered the Kenyan government to regularise the process within one year, and noted that after this deadline, the land would revert once again to the community (Environment and Land Court, 2021, para. 158). After the one-year deadline had passed, the consortium sought an extension from the high court, saying that the local government had not taken any actions to regularise the title deeds and that the company risked losing the property to the community. In May 2023, their application was rejected, and the land is now set to be returned to locals (Andae, 2023).

This decision, also viewed as a landmark ruling, is perceived as offering an opportunity for Indigenous communities to seek compensation. The

Indigenous pastoralist communities regarded the title deeds as an illegal seizure of their land because the land lease was transferred to private investors without following FPIC procedures. Investors, in turn, faced prolonged and expensive delays due to the neglect of Indigenous rights. Even though the court affirmed the rights of the community to participate in the process, the government has not been able to fully rectify the illegal lease since the wind farm is already completed and the community has been displaced.

Gunaá Sicarú Wind Project in Oaxaca, Mexico

Since 2003, when Mexico designated Oaxaca as a key wind-energy development area and initiated incentives for wind-energy projects through climate-change mitigation legislation, there have been numerous prominent conflicts between Indigenous communities in the region and wind-energy developers. The conflicts often concern violations of FPIC, land seizures and other human rights abuses (Dunlap, 2017). Mexican courts have ruled on the issue of FPIC in renewable energy projects on several occasions, but Indigenous community conflicts with wind-energy developers continue (Dudine & Szoke-Burke, 2020). The Gunaá Sicarú development is but one continuation of these conflicts and concerns a plan to build a 300 MW wind farm with 115 turbines on the communal land of the Zapotec Indigenous community of Unión Hidalgo, in Oaxaca. In 2016, a French state-owned company, Électricité de France (EDF), together with its Mexican subsidiary, Eólica de Oaxaca, applied for operational permits and tax exemptions from the local government, carried out an EIA and signed a partnership with the Mexican Federal Electricity Commission. They also began negotiations and signed land contracts for the wind park project with select individuals acting as private landowners (Loyer et al., 2021). In 2017, EDF finalised an energy supply contract and entered into a memorandum of understanding with the Oaxaca state government for the development of a wind farm, all without conducting a proper consultation with the members of the Unión Hidalgo community. The construction permit was granted before consultation with the communal assembly took place, as required by Mexican agrarian law, before the Mexican government conducted a consultation process in accordance with national law and ILO Convention 169, and before a social impact assessment was done, as required by the Mexican law on electrical power (ECCHR, 2020; France NCP, 2020).

Zapotec and other Indigenous communities in the Unión Hidalgo and Juchitán de Zaragoza regions of Oaxaca have claimed that EDF did not secure community FPIC for the wind farm (Azzopardi, 2022), and that electricity companies and local authorities had resorted to 'divide and rule tactics' to obtain approval for the project (Forst, 2018). The Indigenous communities filed a complaint with the OECD National Contact Point (NCP) in France in 2018, asserting that EDF had failed to respect their right to FPIC, and in doing so, had violated the OECD Guidelines (OECD Watch, 2018). The community underwent mediation with EDF for a year, but the community withdrew its claim once it realised that the NCP was not taking the people's concerns seriously and was unable to order temporary cessation of the project (OECD Watch, 2018). Finally, the NCP asserted that EDF 'could have conducted comprehensive due diligence measures on social, cultural, and customary aspects concerning the Unión Hidalgo Indigenous community' and that such measures could have included additional consultations with the Indigenous people. This would have allowed EDF to 'better identify and prevent the societal risks linked to that project' (France NCP, 2020). All along, EDF has refuted that the wind farm project is located on Indigenous ancestral land, asserting that all its infrastructure development is occurring on private land (BHRRC, 2021b).

The representatives of the Unión Hidalgo communities decided to file a claim in the Mexican Federal Court demanding more information about the project and the right to obtain protective measures against any physical harm, while condemning the violation of Mexican and international law regarding FPIC and demanding that the permits be invalidated (Loyer et al., 2021). In 2018, the court ordered a consultation process that must be carried out in accordance with the international standards defined under ILO Convention 169 and the UN Declaration on the Rights of Indigenous Peoples (ECCHR, 2020). All these developments escalated the number of attacks against the Indigenous community members and human rights defenders to the extent that the National Human Rights Commission agreed to issue protective measures for the community (Loyer et al., 2021). In October 2020, the Unión Hidalgo representatives, together with two NGOs, filed a civil lawsuit at the Paris Civil Court against EDF under the French duty of vigilance law (*Indigenous community of Unión Hidalgo, ProDESC and ECCHR v. EDF*). As a French company, EDF is obligated under the duty of vigilance law to identify risks and take measures to prevent and address human rights violations and significant environmental damage related to its operations, as well as those of its subsidiaries, suppliers and subcontractors. Furthermore, given its ownership by the state, the French government also bears a responsibility to address these international human rights.

The claimants argued that EDF had interfered with the FPIC process when implementing its wind park project, thereby breaching its duty of vigilance (BHRRC, 2020). The complaint additionally asserted that EDF's vigilance plan did not adequately identify the risks associated with the Gunaá Sicarú project, lacking specific and comprehensive measures to avert potential violations (BHRRC, 2020). Consequently, the violation of FPIC and lack of approval from the local community had resulted in internal discord, at times intensifying into violent conflicts. EDF responded by dismissing all allegations and stated that their vigilance plan met the requirements of the law (Loyer et al., 2021). In November 2021, the Paris Civil Court recognised its jurisdiction to address the issue but in part dismissed the lawsuit on procedural grounds. The court also declined the claimants' plea to suspend EDF's wind park construction until the community's right to FPIC was guaranteed through the implementation of a new due diligence plan (Ewell, 2022, p. 314). In 2021, a federal court in Oaxaca ordered the suspension of Gunaa Sicarú's construction operations in response to the community's resistance. In June 2022, the Mexican local electricity company terminated its power supply agreement with EDF, resulting in the practical abandonment of the project (ProDESC, 2022). Later that same year, the Ministry of Energy cancelled its supply contract with EDF. The Gunaa Sicarú case, along with the cases from Norway and Kenya, demonstrates how renewable energy companies infringe upon Indigenous and local peoples' rights, but it also proves how Indigenous communities do also have some strength vis-à-vis industries.

Conclusions

Indigenous people are currently disproportionally impacted by clean energy investments due to the land-intensive nature of wind and solar energy. As the case studies show, the dominant understanding of clean investments as 'green' and 'sustainable' is sometimes misleading. The recurrent themes in clean energy projects seem to be the significant impact they have on Indigenous people as well as inadequate acquisition of FPIC consent or insufficient consultation with the Indigenous communities affected by those projects (OHCHR, 2017; Fjellheim, 2023a; Cambou, 2024), which calls into question the sustainability of clean energy projects. The above cases also demonstrate that renewable energy companies play a critical role in determining how Indigenous peoples' rights are upheld in wind energy projects since they wield considerable influence in preventing, mitigating and remedying the adverse human rights impacts stemming from their activities.

Given the growing focus on the need to address climate change and the urgent need for a transition to renewable energy, it is imperative to accelerate finance mobilisation for clean energy projects, but not at the expense of the rights of Indigenous peoples. Some industry sectors, companies and investment funds have incorporated FPIC procedures into their internal guidelines, but as the empirical evidence demonstrates, many corporations do not take FPIC seriously and push only for business-friendly results, which has led to litigation in diverse courts of law against renewable energy companies and their investors or to the filing of complaints through alternative dispute resolution mechanisms. Such instances have generally resulted in a recognition of the evolution of international law on FPIC or have affirmed the significance of community consultation before commencing with the development of renewable energy installations.

The cases presented above demonstrate the legal protection that courts are willing to afford to Indigenous people, despite significant economic investments and political interests being in play. The involvement of corporations and investors in the cases increases the potential for additional rulings on corporate responsibility to uphold Indigenous peoples' rights, particularly rights pertaining to FPIC, challenging traditional views that consider Indigenous rights solely a state obligation (Roesch, 2016). To ensure that corporations abide by their obligation to respect Indigenous rights, it is crucial that corporate FPIC processes complement the state's obligation (Ewell, 2022). As suggested in the existing literature, establishing clearer legal guidelines for corporate FPIC responsibilities and creating accountability mechanisms to better assess and enforce these commitments, instead of relying solely on voluntary measures, could ensure compliance with more robust human rights standards. Mandatory corporate obligations become especially crucial in situations where the state is either unwilling or unable to enforce robust FPIC protections and where community members cannot seek redress through or against the state. Host states could also incorporate, through legal measures, mandatory community participation procedures into investment assessment processes and negotiations regarding investor–state contracts (Agrawal et al., 2023). This type of integration could address human and land rights issues at the initial stage of the project approval cycle.

However, it is also important to acknowledge that forcing corporations to abide by their FPIC obligations is not enough or always the solution for empowering Indigenous communities (Young, 2019; Eisenberg, 2020). Not all of the positive developments within the field of FPIC automatically function as mechanisms for their self-determination or provide Indigenous people with control over a state-based legal system (Young, 2019). Likewise, FPIC does not give Indigenous communities veto power or the capacity to halt all forms of land use. Rather, it means that the project developer must convincingly show that it has actively engaged in dialogue aimed at reaching an agreement with Indigenous communities and has thoroughly evaluated the cumulative impacts of the project on Indigenous culture and livelihoods. Widespread community support, Indigenous leadership, relationships with the 'colonial state' and corporations, as well as the inclusion of local culture and local knowledge are therefore regarded as essential for ensuring the project's sustainability and for attaining a fair distribution of benefits (Lipp & Bale, 2018; Fjellheim, 2023a; Tzay, 2023; Cambou, 2024).

References

Agrawal, H., El-Katiri, L., Muiruri, K., & Szoke-Burke, S. (2023). *Enabling a just transition: Protecting Human Rights in renewable energy projects*. A Briefing for Policymakers. CCSI/Align.

Åhrén, M. (2016). *Indigenous peoples' status in the international legal system*. Oxford University Press.

Andae, G. (2023, May 26). Sh80 billion Lake Turkana wind power loses land to community. *Business Daily*. https://www.businessdailyafrica.com/bd/markets/commodities/sh80-billion-lake-turkana-wind-power-loses-land-to-community%2D%2D4247286

Azzopardi, T. (2022, June 7). Mexico Cancels Contract with Controversial EDF Wind Farm. Wind Power Monthly.

Business & Human Rights Resource Centre (BHRRC). (2020, October 13). EDF lawsuit (re indigenous rights in Mexico, filed in France). https://www.business-humanrights.org/en/latest-news/edf-lawsuit-re-indigenous-rights-in-mexico-filed-in-france/#:~:text=The%20lawsuit%20alleges%20that%20the,The%20case%20is%20ongoing

Business & Human Rights Resource Centre (BHRRC). (2021a, November). Renewable Energy & Human Rights Benchmark: Key Findings from the Wind & Solar Sectors. https://media.business-humanrights.org/media/documents/2021_Renewable_Energy_Benchmark_v5.pdf

Business & Human Rights Resource Centre (BHRRC). (2021b, September 27). Mexico: Court ruling bans wind farm construction on indigenous lands, but EDF denies that it will affect the development of its Gunaa Sicarú project. https://www.business-humanrights.org/en/latest-news/mexico-court-ruling-bans-wind-farm-construction-on-indigenous-lands-but-edf-denies-that-it-will-affect-the-development-of-its-gunaa-sicar%C3%BA-project/

Business & Human Rights Resource Centre (BHRRC). (2024). Just transition litigation tracking tool. https://www.business-humanrights.org/en/from-us/just-transition-litigation-tracking-tool/

Cambou, D. (2020). Uncovering injustices in the green transition: Sámi Rights in the development of wind energy in Sweden. *Arctic Review on Law and Politics, 11*, 310.

Cambou, D. (2024). The significance of the Fosen decision for protecting the cultural Rights of the Sámi indigenous people in the green transition'. In D. Cambou & Ø. Ravna (Eds.), *The significance of Sámi rights. Law, justice, and sustainability for indigenous Sámi in the Nordic countries.* Routledge.

Danwatch. (2022). Violations of human rights are delaying the green transition in Kenya. https://danwatch.dk/en/perspektiv/brud-pa-menneskerettigheder-bremser-den-gronne-omstilling-i-kenya/

Deva, D. (2012). *Regulating corporate Human Rights violations: Humanizing Business.* Routledge.

Dolton-Zborowski, S., & Szoke-Burke, S. (2022). *Respecting the Human Rights of communities: A legal risk primer for commercial wind and solar project deployment.* CCSI.

Dunlap, A. (2017). 'The town is surrounded:' From climate concerns to life under wind turbines in La Ventosa, Mexico. *Human Geography, 10*(2). https://doi.org/10.1177/194277861701000020

Dunlap, A. (2018). A bureaucratic trap: Free, prior and informed consent (FPIC) and wind energy development in Juchatan, Mexico. *Capitalism Nature Socialism, 29*, 1–21.

Eisenberg, A. (2020). Consent, resistance and the duty to consult. *International Journal on Minority and Group Rights, 27*, 270–290.

Environment and Land Court. (2021, October 19). Judgment. Meru ELC case 163 of 2014.

Equator Principles. (2020, July). EP4.

Equitable Origin. (2024). EO100™ Standard for responsible energy development.

European Centre for Constitutional and Human Rights (ECCHR). (2020). Case Report: Wind farm in Mexico: French energy firm EDF disregards indigenous rights.

Ewell, C. (2022). Reimagining the renewable energy transition: The potential for mandatory corporate due diligence to ensure respect for the right to free, prior, and informed consent. *Yale Journal of International Law, 47*, 309–344.

Finnfund. (2023, March 16). Finnfund sells its share in Lake Turkana Wind Power to BlackRock. *Finnfund*. https://www.finnfund.fi/en/news/finnfund-sells-its-shares-in-lake-turkana-wind-power-to-blackrock/.

Fjellheim, E. (2023a). Wind energy on trial in Saepmie: Epistemic controversies and strategic ignorance in Norway's green energy transition. *Arctic Review on Law and Politics, 14*, 140–168.

Fjellheim, E. (2023b). "You can kill us with dialogue:" critical perspectives on wind energy development in a Nordic-Saami green colonial context. *Human Rights Review, 24*, 25–51.

Forst, M. (2018). Report of the Special Rapporteur on the situation of human rights defenders on his mission to Mexico. A/HRC/37/51/Add.2.

IRENA. (2023). *World energy transitions outlook 2023: 1.5°C pathway*. IRENA.

Kårtveit, B. (2021). Green colonialism: The story of wind power in Sápmi. In R. Sørly, T. Ghaye, & B. Kårtveit (Eds.), *Stories of change and sustainability in the Arctic regions. Interdependence of local and global*. Routledge.

Knutsson, P., & Ibarlucea Flores, P. (2022). *Trends, investor types and drivers of renewable energy FDI*. OECD Working Papers.

Lehr, A., & Smith, G. (2010). *Implementing a corporate free, prior, and informed consent policy: Benefits and challenges*. Foley Hoag eBook.

Lipp, J., & Bale, S. (2018, February). *Growing indigenous power: A review of Indigenous Involvement & Resources to further renewable energy development across Canada*. TREC.

Lomax, J., Mirumachi, N., & Hautsch, M. (2023). Does renewable energy affect violent conflict? Exploring social opposition and injustice in the struggle over the Lake Turkana wind farm, Kenya. *Energy Research & Social Science, 100*, 103089.

Loyer, C., Bommier, B., Lavite, C., & Torres, G. (2021). Vigilance switched off: Human Rights violations in Mexico: What are the responsibilities of EDF and the APE? Report, June 2021.

Muchlinski, P. (2014). Corporations in international law. In *Max Planck Encyclopaedia of International Law*. Oxford University Press.

Normann, S. (2020). Green colonialism in the Nordic context: Exploring southern Saami representations of wind energy development. *Journal of Community Psychology, 49*(1), 77–94.

O'Neill, L., Thorburn, K., Riley, B., Maynard, G., Shirlow, E., & Hunt, J. (2021). Renewable energy development on the indigenous estate: Free, prior and informed consent and best practice in agreement-making in Australia. *Energy Research & Social Science, 81*. https://doi.org/10.1016/j.erss.2021.102252

OECD Watch. (2018). Union Hidalgo vs EDF Group. EDF's violation of FPIC at wind energy park in Mexico.

OECD Watch. (2020). Society for Threatened Peoples Switzerland vs. BKW group. BKW Energie AG'S investment in wind production on Sami territory, Norway.

OHCHR. (2017, April 27). Report of the Working Group on the issue of human rights and transnational corporations and other business enterprises on its mission to Mexico. A/HRC/35/32/Add.2.

Perrone, N. (2022). Local communities, Extractivism and international investment law: The case of five Colombian communities. *Globalizations, 19*(6), 837–853.

ProDESC. (2022, June 2). Setback to EDF in Mexico! The Zapotec Community of Unión Hidalgo and ProDESC Achieve the Definitive Cancellation of the Wind Park Megaproject Gunaa Sicarú. https://prodesc.org.mx/en/set-back-to-edf-

in-mexico-the-zapotec-community-of-union-hidalgo-and-prodesc-achieve-the-definitive-cancellation-of-the-wind-park-megaproject-gunaa-sicaru/
Rambouts, S. J. (2014). *Having a say: Indigenous peoples, international law and free, prior and informed consent.* Doctoral Thesis, Tilburg University.
Rasmussen, E. (2023, December 13). Norway to pay Sámi reindeer herders millions. *ICT.* https://ictnews.org/news/norway-to-pay-s%C3%A1mi-reindeer-herders-millions
Ravna, O. (2023). The Fosen case and the protection of Sámi culture in Norway pursuant to article 27 ICCPR. *International Journal on Minority and Group Rights, 30,* 1–20.
Renewables Now. (2020, January 20). New complaint says 288-MW Norwegian wind farm violated Sami rights. https://renewablesnow.com/news/new-complaint-says-288-mw-norwegian-wind-farm-violated-sami-rights-684075/
Roesch, R. (2016). The story of a legal transplant: The right to free, prior and informed consent in sub-Saharan Africa. *African Human Rights Law Journal, 16,* 505–531.
Saami Council. (2023, October 12). Statkraft violates its self-imposed human rights obligations. https://www.saamicouncil.net/news-archive/statkraft-violates-its-self-imposed-human-rights-obligations
Schilling-Vacaflor, A., & Flemmer, R. (2020). Mobilising free, prior and informed consent (FPIC) from below: A typology of indigenous peoples' agency. *International Journal on Minority and Group Rights, 27,* 291–313.
Smith, A., & Scott, D. (2022). Law, labour and landscape in a just transition. In U. Natarajan & J. Dehm (Eds.), *Locating nature.* Cambridge University Press.
Society for Threatened People. (2023). Norway: Wind power threatens south Sámi peoples' reindeer herding. https://www.gfbv.ch/en/publications/case-studies/fosen-vind/
Society for Threatened People vs. BKW. (2020). Specific Instance 16.1.2020.
Special Rapporteur. (2016). Report of the special Rapporteur on the rights of indigenous peoples. UN Doc A/HCR/33/42.
Statkraft. (2024). Fosen vind. https://www.statkraft.com/about-statkraft/where-we-operate/norway/fosen-vind/
Stuart, A., Bond, A., Franco, A., Baker, J., Gerrard, C., Danino, V., & Jones, K. (2023). Conceptualising social licence to operate. *Resources Policy, 85,* 103962.
Supreme Court of Norway. (2021). Judgment. HR-2021-1975-S.
Tzay, J. (2023). *Green financing—A just transition to protect the rights of indigenous peoples.* UN Doc A/HRC/54/31.
UN Global Compact. (2004). The Ten Principles of the UN Global Compact.
UNCTAD. (2023). *World investment report 2023.* UN.
Union Hidalgo vs EDF Group. (n.d.). EDF'S violation of FPIC at wind energy park in Mexico. *Date filed, 8*(2), 2018.
Williams, S. (2023). Understanding indigenous rights risks in renewables. Schroders May 2023.
Young, S. (2019). *Indigenous peoples, consent and Rights: Troubling subjects.* Routledge.
Zografos, C. (2020). The contradictions of green new deals: Green sacrifice and colonialism. *Soundings, 80,* 37–50.

CHAPTER 6

Exploring Sustainability, Indigenous Rights, and Decoloniality: A Content Analysis of Human Rights Language in SDG Reports

Johanna Ohlsson

Introduction

The relationship between sustainability, Indigenous rights, and decoloniality theory occupies critical space in the global pursuit of a more sustainable future. The concept of sustainable development brings together various notions that are not necessarily easily combined based on their different underlying assumptions. The discourse surrounding the United Nations Sustainable Development Goals (SDGs) has, according to Markus Kaltenborn et al. (2020), emerged as a focal point for understanding human rights' role, function, and implementation. For instance, the interconnectedness of human rights and the SDGs is nowadays often taken for granted, as, intuitively, sustainable development and human rights should mutually reinforce each other (Kaltenborn et al., 2020, p. 2). Kaltenborn et al. have argued that the predominant discourse on human rights is often stated as foundational to the SDGs. Also, they state that it serves as the baseline for any SDG implementation, implying that all SDG-related work should have a rights-based dimension to it (Kaltenborn et al., 2020, p. v). For instance, the preamble of the *Transforming Our World: 2030 Agenda for Sustainable Development* explicitly states that "the 17 Sustainable Development Goals [...] seek to realize the human rights of all" (UN General Assembly, UNGA, 2015) and the United Nations Office of the High Commissioner for Human Rights (UN OHCHR) states that "OHCHR has made a strong contribution to the integration of human rights throughout the process to

J. Ohlsson (✉)
Department of Human Rights and Democracy, University College Stockholm, Bromma, Sweden
e-mail: johanna.ohlsson@ehs.se

R. Toivanen et al. (eds.), *Decolonizing the Sustainable Development Goals*,
https://doi.org/10.1007/978-3-032-14263-4_6

define the 2030 Agenda and its SDGs and seeks to ensure that strategies and policies to implement the 2030 Agenda are human rights-based and leave no one behind", stressing equality and non-discrimination i.e. the cornerstones of human rights, as foundational aspects for the SDGs (UN OHCHR, 2024).

So, why can this seemingly straightforward relationship be problematic, and why are the connections between Indigenous rights and the SDGs seemingly at unease? Previous research suggests at least two potential explanations. The first one is related to the implementation gap commonly seen in human rights work, which "attribute[s] gaps between human rights principles and practices to wilful disobedience, self-interested defection, and ineffective enforcement" (Cole, 2015, p. 405). A second explanation focuses on the fact that the language and terminology are insufficient. This was brought forward by disappointed human rights advocates when the *Transforming Our World: The 2030 Agenda for Sustainable Development* (UNGA, 2015) and the SDGs were launched, as "the SDGs had not reflected the advice provided by global leaders and grassroots activists to keep human rights central to the new development era" (Winkler & Williams, 2017, p. 1023). In this chapter, I explore the human rights language in the SDG discourse by analysing the UN's annual progress reports.

Despite the connection, human rights implementation and sustainable development have not always been seen as intertwined processes. As discussed by Kaltenborn et al., Professor and UN Special Adviser Philip Alston highlighted already in 2005 — then concerning the SDGs predecessors, the Millennium Development Goals (MDGs) — the siloed discussions and processes and the lack of coordination between human rights and development communities as a main challenge (Alston, 2005, pp. 757ff.; Kaltenborn et al., 2020, pp. 1ff.). Important steps have been taken to bridge this siloed approach, and developments of the SDG discourse have contributed towards this, partly as they are more inclusive and holistic than the MDGs. The attempts to bring the discourses and practices closer have been commendable, and the SDGs are integrated into governance structures across the globe. However, when the SDGs, their targets, and indicators are implemented, there is a risk that they do so at the expense of human rights implementation, as the siloed approach is still present and continues to pose challenges. For instance, some SDGs (i.e. SDG 7 (Affordable and clean energy), SDG 9 (Industry, innovation, and infrastructure), SDG 11 (Sustainable cities and communities), SDG 13 (Climate action), and SDG 15 (Life on land)) tend to be siloed, i.e. separated from human rights language (Muñoz Carmona, 2024, pp. 80 f.). Further, I argue that replacing the human rights discourse with the SDG discourse risks undermining human rights implementation.

From an Indigenous and minority point of view, scepticism towards the role and use of human rights discourse as well as the SDG discourse (i.e. guiding documents, policies as well as policy and scholarly discussions) is not surprising, as human rights have repeatedly been misused in the pursuit of utilising power structures and relations, reinforcing power imbalances rather than challenging them (Namli, 2020). Yet, this is not necessarily only a question of asymmetrical power relations, but rather about striving for social justice. That rights are used/misused in several ways — even as far as repressive mechanisms — is undermining the very legitimacy of the claims, and has been shown in various contexts, often connected to racialised stereotypes (Namli, 2020). This goes in tandem with decoloniality and postcolonial scholarship (cf. Mutua, 2002; Grovogui, 2006; Sultana, 2018; Struckmann, 2018; Gómez Sánchez, 2020; Arora-Jonsson, 2023) which emphasises that the colonial order is in no way a thing of the past but very much present and continuously reproduced in current structures. One of the central problems within the combination of the SDGs and human rights discourses is echoed in the work of Celermajer and Lefebvre, who state that to "deploy the tools of this system to remedy their situation, then, Indigenous peoples are forced to turn to those responsible for the ongoing violations of their rights" (2020, p. 13). This is indicative of skewness and asymmetry.

The intuitive interconnectedness of human rights and the SDGs rather adds a double jeopardy problem, as Indigenous groups risk facing the misuse of both the human rights and the SDG discourse. Additionally, scholars have argued that the SDG discourse has fundamentally failed to consider other forms of development, hiding certain norms and power structures within the SDGs (Struckmann, 2018; Esquivel, 2016). Also, Winkler and Satterthwaite (2017) have warned of persistent inequalities in the SDGs, and, in tandem with repeated warnings of generalised notions of sustainability, voice concerns over risks overshadowing and suppressing any perspective that differs from the mainstream discourse, including Indigenous perspectives (Toivanen et al., 2026).

The starting point here is the connections between the SDGs and human rights discourses. By addressing two central tensions, *political compromise* vs *philosophical understanding* and the *legal standing* of the instruments in the discourses, through the lenses of Indigeneity and decolonial critique, the aim is to explore if and how Indigenous rights are an integral part of enforcing the SDGs, without undermining the Indigenous rights discourse. This chapter speaks to previous literature in combining critical human and Indigenous rights as well as sustainability studies. It contributes an analysis of the role and function of human and Indigenous rights in the SDG reports from 2016 to 2023, stressing the centrality of a decolonial perspective. The material is selected based on its positionality within the UN system and the annual reports' focus on summarising how the implementation of the SDGs is progressing.[1]

[1] Please note that the annual reports are different from the Secretary-General's progress reports to the UN General Assembly's Economic and Social Council, which also are published yearly by the UN. The selected material offers a complete selection, as all reports since the SDGs were initiated are included. This makes eight reports, ranging from 40 to 80 pages yet following a similar format. The material is discussed below.

The critical contribution of this chapter is that of an ethicist, to take a step back, identify the normative assumptions underlying political and pragmatic practices, and critically discuss the moral conditions at hand.

Human Rights, Sustainability, and the Tensions In Between

The human rights discourse is here seen as a broader umbrella where various kinds of subcategories are situated, and Indigenous rights are straightforwardly seen as human rights. Human and therefore also Indigenous rights are legitimate claims individual rights-holders have towards the corresponding duty-bearer. In the context of human and Indigenous rights, the concepts of duty-bearers and rights-holders are pivotal for understanding the dynamics of responsibility, but also power structures. Rights-holders are individuals or groups who inherently have certain rights, irrespective of their background or identity (Hohfeld, 1964). These rights, often considered universal, include aspects such as the right to life and protection against discrimination.

Duty-bearers, on the other hand, are entities or individuals with the responsibility to respect and protect rights, commonly the state. Governments, as the prominent duty-bearer, are expected to create and enforce laws that safeguard human rights, ensuring that the rights-holders within their jurisdictions are treated with dignity. This is governed by a system of human rights instruments, and the pioneering Universal Declaration for Human Rights (UDHR) of 1948, the two core conventions from 1966 (in force 1976: the Convention on Economic, Social and Cultural Rights (ICESCR) and the Convention on Civil and Political Rights (ICCPR)) as well as the following themed conventions are all to be seen as supplementary legal and normative mechanisms which together make up the Catalogue of Rights. However, these instruments have various legal standings. The conventions are binding under international law once enough states have ratified them, while declarations have a non-binding legal status, even though they are part of soft law (cf. Hannum, 1995; Shaw, 2021), such as the United Nations Declaration on the Rights of Indigenous Peoples (UNDRIP).

Indigenous rights are commonly seen as the subcategory of rights that are inherent to Indigenous peoples. These rights have deep historical roots, shaped by centuries of colonisation, discrimination, and dispossession. Indigenous peoples have the right to the full enjoyment, as a collective or as individuals, of all human rights and fundamental freedoms as recognised in the Catalogue of Rights, i.e. international human rights law.[2]

[2] Several of the core conventions in the Catalogue of Rights address groups of people who have been identified as particularly exposed or vulnerable to aspects of discrimination and abuse of power (such as the Convention of the Rights of the Child, CRC, the Convention on the Elimination of All Forms of Discrimination against Women, CEDAW, and the Convention on the Rights of Persons with Disabilities, CRPD).

UNDRIP has been a valuable tool for the striving towards recognition and self-determination for Indigenous peoples. It is "the first universal instrument to recognise and elaborate the content of the right to self-determination for indigenous peoples. It is also the first instrument to recognize the right of self-determination for peoples other than peoples territorially organized as states and colonies" (Cambou, 2019, pp. 34–35). When adopted by the UNGA in 2007, 143 states voted in favour, 11 abstained, and four states voted against it. The four states voting against were Australia, Canada, New Zealand, and the United States, notably states with Indigenous peoples amongst their population. Some years later, these four countries reversed their position and now support UNDRIP. The 11 abstentions in 2007 were Azerbaijan, Bangladesh, Bhutan, Burundi, Colombia, Georgia, Kenya, Nigeria, the Russian Federation, Samoa, and Ukraine (UN DESA, 2007). Yet, the mechanisms where UNDRIP is the most central, specifically directed towards Indigenous rights, also have various legal standings, and the mechanisms of Indigenous rights vary. UNDRIP is "only" a declaration and not a convention, giving it a soft law status, even "softer" than the international conventions on human rights from a national law perspective. On the other hand, given that all other human rights instruments also apply, there are various tools at hand for implementing and making sure that decision-making complies with — instead of failing to meet — the human rights obligations.

This discussion highlights one tension this chapter addresses, i.e. the various legal standings. We see that different human rights instruments have varying legal standings, and this is the case also for the SDGs. Yet, despite Agenda 2030 being primarily a political compromise, it also has legal implications. Legal scholars such as Österdahl have shown that the UN General Assembly resolution *Transforming our World: the 2030 Agenda for Sustainable Development* is a non-legally binding instrument from which the SDGs are developed, which still can affect states' behaviours (Österdahl, 2023).

When it comes to the discourse of sustainable development, the baseline is commonsensically an understanding of development that is sustainable. This leads to a conceptual focus on how to understand (a) development and (b) sustainability. The policy-oriented discourse where sustainable development is positioned often takes the Brundtland Report (in short) or the World Commission on Environment and Development (WCED) publication *Our Common Future* (1987) as a starting point. The WCED states that "Sustainable development is development that meets the needs of the present without compromising the ability of future generations to meet their own needs" (WCED, 1987, p. 43). As Grenholm (2007) and others identify, this includes an understanding of human responsibility towards future generations, an obligation to reduce poverty by promoting economic development, and responsibility towards the environment and conservation of nature (Grenholm, 2007, p. 9). By bringing the development and the environmental debate closer together, the Brundtland Report has generated ample discussions on the meaning, connections, and implications of the two, and has served as a successful agenda-setting exercise (McNeill, 2007, p. 24). For instance, this discussion has brought attention to several questions of moral and political character such as (1) which subjects have moral (and legal) standing, and to what extent humans and non-humans have rights, but also (2) how distribution of wealth and resources should be understood and to what extent this relates to models of distributive justice, and (3) how the rights of future generations should be understood, in particular in relation to current injustices (McNeill, 2007, p. 24).

Sustainable development is in the currently established tripartite concept covering the dimensions of social, economic, and environmental sustainability (Purvis et al., 2019). This, according to the UN, involves juggling "economic growth, social inclusion, and environmental protection" as all three aspects are equally important, and "where development is sustainable, everyone has access to decent work, quality health care, and education. Natural resource use avoids pollution and permanent losses to the environment. Public policy choices ensure that no one is left behind due to disadvantages or discrimination" (UN Sustainable Development Summit, 2023). Yet, the critical reader will remember that these wordings result from a political process making them a compromise, and of a rhetorical character. A critical analysis of the concept and, perhaps less generously, the buzzword that sustainability has become indicates that it is fairly empty and needs to borrow normative content from, e.g., justice and well-being. Hence, sustainability sets out to align with the constitutional and international law principle of dignity (Herlin-Karnell, 2023), but also more generally with principles of equality, freedom, and non-discrimination.

Rights in the SDG Discourse

The use of human rights terminology offers a couple of avenues for exploration within the SDG discourse. For instance, Kaltenborn, Krajewski, and Kuhn have highlighted how the majority of the SDGs' objectives, concrete goals, targets, and indicators correspond to human rights obligations (2020, p. 2). Findings from the Danish Institute for Human Rights' mapping of the SDGs show "that 156 of its 169 targets (more than 92%) reflect human rights and basic labour standards" (Feiring & Hassler, 2016). In this mapping, the SDGs are connected to specific articles in various human rights instruments such as the ICESCR, ICCPR, the Convention of the Rights of the Child (CRC), the Convention on the Elimination of All Forms of Discrimination against Women (CEDAW), and various protocols and general comments to the conventions. There are also connections identified to mechanisms from other international organisations, such as the International Labour Organization (ILO). Compliance with UNDRIP as well as the other central mechanisms within international law such as the International Labour Organization's Indigenous and Tribal Peoples Convention No. 169 is a key aspect in the support of Indigenous peoples rights. Importantly, given that Indigenous rights are integral to human rights, all other instruments are also legally valid and significant in practice.

The connections identified between SDGs and human rights discourses in previous works can be interpreted more or less generously. A generous interpretation focuses on these connections and the reinforcing potential they offer, while a less generous interpretation identifies this as simply rhetorical. This chapter acknowledges that the connections are there, but they are weak and implicit, which leads to a requirement for further analytical work. Also, this acknowledges the risk that human rights are reduced to rhetoric within the SDG discourse.

I argue that this connection is at best dependent on a generous interpretation because the implementation of human rights is not the same as terminological connections per se. Simply mentioning human rights in policy documents does not lead to their implementation. Instead, it risks creating a false imaginary, where addressing various SDGs and targets automatically equals implementing human rights. Therefore, a critical understanding of the connections between human rights and the SDGs must do more than acknowledge these connections; it must critically review and analyse them. Acknowledgement is the first step in the process of implementing human rights within the procedure of the SDGs, but it can also be entangled with various simplifying assumptions.

Mapping the use of human rights terminology through quantitative and qualitative text analysis offers a systematic view of the discursive landscape at hand. The frequency at which the terms are used in key texts serves as an indicator of their importance; however, this must be coupled with a more substantial, qualitative analysis of *how* the terminology is used. In addition, critical human rights studies offer a platform for scrutinising the assumed interconnectedness between SDGs and human rights in the general sense. For instance, a systematic assessment identifying the most common themes (such as children's or Indigenous rights) also tells us something about the prioritisation and framing(s) of certain issues.

This chapter argues that the connection between human rights and the SDGs is particularly significant in the context of Indigenous rights, as historical injustices and systemic inequalities have often resulted in the violation of the rights of Indigenous communities. Also, as Corntassel (2008) has argued, the definition of sustainable development in the Brundtland Report from 1987

> ... does not go far enough as a benchmark for indigenous political, cultural, economic, and environmental restorative justice (in theory and in practice). For indigenous peoples, sustainability is intrinsically linked to the transmission of traditional knowledge and cultural practices to future generations. Without the ability of community members to continuously renew their relationships with the natural world (i.e., gathering medicines, hunting and fishing, basket-making, etc.), indigenous languages, traditional teachings, family structures, and livelihoods of that community are all jeopardized. (Corntassel, 2008, p. 118)

This caveat of the definition stresses the necessity of assessing the SDGs through Indigenous and decoloniality perspectives.

Critique of and Reproductions in the Discourses

The connections between sustainability, Indigenous rights, and decoloniality theory generate a complex web highlighting the interconnected nature of social, economic, environmental, and cultural considerations. Scholars using sustainability as a concept commonly seek to address the well-being of present and future generations by promoting equity, justice, and inclusivity (Agyeman et al., 2002; Agyeman, 2013). Further, as Toivanen and Cambou have argued, human rights are key concepts of sustainability science because they constitute the basis for sustainable well-being in any given society (Toivanen & Cambou, 2021). On the other hand, previous literature also shows that human rights and development do not always go hand in hand (Woods, 2010), even less so with Indigenous rights, as the SDGs have little to offer to the agency of Indigenous peoples. For instance, Toivanen and Cambou have shown that when Indigenous rights are discussed, it is done in a manner that implies victimhood and vulnerability rather than agency and representation (Toivanen & Cambou, 2021).

Recognising and respecting Indigenous rights is paramount for many reasons, but in particular from a decolonial perspective. However, human rights have also received their fair share of post- and decolonial critique, due to their embedded power structures, which more often correspond to and reinforce, instead of challenging, problematic power hierarchies (cf. Mutua, 2002; de Jong, 2022). Also, Indigenous and postcolonial critics have emphasised and problematised the process of othering non-whiteness and Indigeneity in relation to the subject of human rights, not only based on prejudicial and racial structures but also stressing the necessity to understand this in the larger picture of imperialism and colonialism (Celermajer & Lefebvre, 2020, p. 8).

Indigenous communities are not merely stakeholders in modern development but also survivors of historical and current injustices. Centrally, individuals within Indigenous groups are subjects with rights, despite the racialised, imperialist, and colonial structures that deem "Indigenous and colonized peoples as unfit to be full legal persona and political subjects". Furthermore, the "historical refusal to recognize Indigenous peoples as full subjects persist in their ongoing exclusion as sovereign peoples in their own lands and the contemporary international order" (Celermajer & Lefebvre, 2020, p. 9). Indigeneity continuously shapes and generates cultural diversity and environmental stewardship, and this must — if rights are to be taken seriously by all duty-bearers (i.e. governments and governmental employers at various administrative levels, policymakers and decision-makers, as well as scholars) — be acknowledged and restored, in a genuine, decolonial way. Indigenous communities have valuable traditional knowledge and practices that bring much-needed nuances to sustainable development efforts worldwide, but this potential is rarely taken into consideration by those with agenda-setting power.

Importantly, key UN documents do not convey the obligations of human rights in a sufficient way (Williams & Blaiklock, 2016). I argue that this leads to confusion about who the duty-bearers are when sustainability is used interchangeably with human rights. In addition, previous research has shown that the SDGs promote a false sense of success and make it easy for governments to work slowly on the realisation of human rights. Even more troublesome, the SDGs fail to specify what a human-rights-based duty or genuine goal to eradicate severe poverty requires: a clear division of labour and the full realisation of human rights requires a massive roll-back of international and intra-national inequalities, which the SDGs fail to demand (Pogge & Sengupta, 2016).

Scholars positioned in a critical discourse commonly ask *who* benefits, but also what kinds of power hierarchies are at play. Current global structures are designed to reinforce colonial power hierarchies, leading to marginalised groups being confined in marginalisation. The discussions in this chapter are informed by a decoloniality perspective, stressing critique against colonial domination by building on the perspectives on decoloniality offered by Mignolo and Walsh (2018). Importantly, de- and postcolonial critique has repeatedly been directed towards the SDGs in general (cf. Arora-Jonsson, 2023; Sultana, 2018; Struckmann, 2018), human rights in particular (cf. Gómez Sánchez, 2020; Grovogui, 2006), as well as the more specific focus on the human rights-based foundation of the SDGs (cf. Arora-Jonsson, 2023). These assessments offer a welcome space and possibility to resume the reconstructions of the normative assumptions on which the human rights and SDGs discourses rest.

While there is potential for constructive interaction, tensions have historically arisen between the ideals of sustainability and Indigenous rights. Conflicts over land rights, resource extraction, and cultural preservation have often hindered progress towards sustainability based on decoloniality. As Pathak Broome, Bajpai, and Shende have identified, "...the mechanism envisioned for achieving the SD goals is in direct conflict with the goals themselves" as there are "three fundamental (among other) lacunae in the mechanism that has been envisioned to achieve SDGs" (2020, p. 68). The first lacuna is that the SDGs rely on the existing economic order and model of economic growth "which are the very fundamental basis for the root causes leading to the very injustices and inequities that the SDGs intend to eliminate" (2020, pp. 68 f.), reinforcing patriarchal, cast, and class-based divides as well as communal divisions. Second, the implementation of SDGs is dependent on states, and top-down processes are predominant. The third lacuna focuses on the "absence of clarity on what basic set of ethical values or principles would underlie the processes in place to achieve all goals, targets, and activities within the SDGs" (2020, p. 69). It is important to recognise that Indigenous communities have a key role in achieving development that aligns with Indigenous value systems and that Indigenous peoples have headed initiatives that contribute positively to their communities as well as broader aspects of SDGs. Indigenous knowledge and practices have repeatedly shown important potential for taking the lead in working towards a more sustainable future (cf. Breidlid & Krøvel, 2020).

The intersectionality arises from the recognition that sustainability cannot be achieved without addressing the specific challenges faced by Indigenous communities. This includes safeguarding Indigenous land rights, preserving cultural heritage, and empowering Indigenous peoples to actively take part in decision-making processes that affect their lives. It requires acknowledging the unique relationship most Indigenous peoples have with their environments and incorporating traditional ecological knowledge into broader sustainability efforts. Furthermore, decoloniality theory prompts a critical examination of existing power structures, urging a move beyond a Eurocentric and West-centric worldview. It therefore challenges the dominant narratives and encourages a more inclusive understanding of sustainability that respects diverse ways of knowing and living.

Themes in Terminology

The analysis of this chapter is based on three parts. The first consists of a quantitative textual analysis of the material (the SDG reports; see details in Table 6.1).

The second part introduces a qualitative analysis of the 2023 report, aiming for an in-depth understanding of the use of human and Indigenous rights. The

Table 6.1 SDG reports 2016–2023

About the report: *The Sustainable Development Goals Report is the UN's official report that monitors global progress on the 2030 Agenda for Sustainable Development* *Using the latest available data and estimates, the report provides a comprehensive assessment of the 2030 Agenda, highlighting not only impacts of multiple crises affecting people's lives and livelihoods, but also areas of progress where acceleration is needed. This annual SDG Report is prepared by the United Nations Department of Economic and Social Affairs (UN DESA), in collaboration with the entire UN Statistical System, consisting of more than 50 international and regional agencies, based on data from over 200 countries and territories*
Reference
UN DESA (2023) *The Sustainable Development Goals Report 2023: Special Edition*. New York, Geneva: United Nations
UN DESA (2022). *The Sustainable Development Goals Report 2022*. New York, Geneva: United Nations
UN DESA (2021). *The Sustainable Development Goals Report 2021*. New York, Geneva: United Nations
UN DESA (2020). *The Sustainable Development Goals Report 2020*. New York, Geneva: United Nations
UN DESA (2019). *The Sustainable Development Goals Report 2019*. New York, Geneva: United Nations
UN DESA (2018). *The Sustainable Development Goals Report 2018*. New York, Geneva: United Nations
UN DESA (2017). *The Sustainable Development Goals Report 2017*. New York, Geneva: United Nations
UN DESA (2016). *The Sustainable Development Goals Report 2026*. New York, Geneva: United Nations

third is a critical analysis of the relationship between human and Indigenous rights and the SDG discourse at large, and what the findings from the content analyses can tell us about the role of human and Indigenous rights in relation to sustainability. The process is divided into these three steps as they jointly build on and develop the argument of the role of human and Indigenous rights in the SDG discourse cumulatively. However, the three-step approach also allows for critical discussion of each step. The first step, the quantitative content analysis, indicates how often rights are mentioned in the texts (see Table 6.2). Given that human rights are seen as the baseline for all SDG-related work (UN OHCHR, 2024; UNGA, 2015), it is surprising that rights are only mentioned 36 times in the most recent report, covering an 80-page document (UN DESA, 2023). This is also the report with the highest frequency. On the other hand, interpretative problems commonly arise when content analyses are conducted, as the deductive method is dependent on choices forming the categories and search words (Esaiasson et al., 2012, p. 197). The search words used are (1) *rights*, (2) *human rights*, and (3) *Indigenous rights*, justified by their centrality to get a better idea of how often the direct rights terminology was used. This categorisation generates a certain overlap in the material, as all searches in the first category would also be included in the second and third. However, as shown in the summary of the results in Table 6.2, *Indigenous rights* was only used once in the most recent report (UN DESA, 2023), while *Indigenous peoples* occurred three times.

The findings from the quantitative content analysis show a need to take further steps to generate more nuanced results. This justifies the second step, the qualitative content analysis, which aims to identify underlying themes. This builds on the findings from the quantitative content analysis and analyses the sections of the reports where *rights*, *human rights*, and *Indigenous*

Table 6.2 Results from quantitative content analysis of the SDG Annual Reports

Year of report	*# of pages*	*# Mentioning of rights*	*# Mentioning of human rights*	*# Mentioning of Indigenous peoples (Indigenous rights)*
2016	56	13	6	2
2017	64	19	8	2
2018	40	11	6	0
2019	64	28	13	1
2020	68	21	11	2
2021	68	27	18	1
2022	68	22	13	1
2023	80	36	10	3 (1)

Table 6.3 Themes in the SDG Annual Reports

Most common theme	*# of times mentioned*
Women's rights, including women's land rights	11
An overarching understanding of the importance of human rights	9
Sexual and reproductive rights	2
The rights of future generations	2
Rights in relation to war, armed conflict, and migration	1
Rights for people with disabilities	1
Indigenous rights (indigenous peoples twice)	1 (2)

peoples are mentioned. A qualitative analysis of the 2023 report[3] shows that out of the 36 times that *rights* is used in the document, only 30 are relevant to this chapter.[4] When references are made to *rights* in an overarching context, the analysis makes it clear that it is still human rights that are referred to (Table 6.3).

The qualitative text analysis also shows that the mentions are either a description of a context (19) or a prompt/recommendation (11), according to my coding. This makes it possible to draw conclusions. First, these reports may not be the primary platform for assessing the use of human rights terminology in the SDG discourse, despite their centrality. Second, and importantly, the use, or rather non-use, of Indigenous rights in these reports suggests two interpretations. First, it does not seem to be relevant to single out Indigenous rights as they are argued to be included in all other human rights mentions too. Second, on the other hand, given the clear connections between sustainability and Indigeneity argued for in previous research, the lack of concrete and direct discussions is surprising.

Conclusion

This chapter has focused on the relationship between the SDGs as well as human and Indigenous rights. More explicitly, it has discussed the preconditions for these connections by focusing on two tensions between the human rights and the SDG discourses (i.e. *political compromise* vs *philosophical understanding* as well as the *varying legal standings*) through the lenses of Indigeneity and decolonial critique.

[3]The second step is focusing on the 2023 report, as this is the most recent but also the most extensive report (in terms of page numbers and breadth). It is the report where the connections between rights and the SDGs hypothetically should be most developed as the work of bridging the silo discussions has had the most time to materialize.

[4]The others refer to other legal rights, a list of abbreviations, or to copyrights.

The analysis indicates that there is a tendency to equal human rights to social sustainability within the SDG discourse, reducing the space, role, and critical function but also potential of human rights. This challenges the understanding and the scope of human rights mechanisms' and instruments' potential as drivers for social change. When rights are not made explicit within the SDG discourse, they lose their critical potential to spur social change. They become simply "nonsense on stilts" (Bentham, 1843) when misused and reduced to rhetoric. In tandem with being reduced to rhetoric, they are reduced to pragmatic tools that are frequently misused, making the postcolonial critique of human rights very much valid. So, yes, when having a reductivist understanding and approach towards rights — whether it is reduced based on its use (i.e. only rhetoric) or on its position within the liberal system (risking contributing to reproducing harms, discrimination, and injustices) — they are no more than nonsense on stilts, but when taken seriously, rights have the potential to serve as critical tools for social change.

The descriptive parts of the chapter have shown that the rights language in the SDG reports is relatively sparse. The quantitative content analysis frames articulations of rights based on three search strings following the aim of this chapter, (1) rights, (2) human rights, and (3) Indigenous rights, exposing a limited use of the rights discourse, generating a critical discussion on the use of terminology. This limited presence is echoed yet nuanced in the qualitative content analysis, where the findings identify four themes as most prominent: (a) an overarching understanding of the importance of human rights, (b) women's rights, including women's land rights, (c) sexual and reproductive rights, and (d) the rights of future generations. The normative parts of the chapter centre on discussions of the underlying assumptions and normative conditions and address a discussion of what wordings and language can tell us. This combination of a descriptive and normative approach leads to findings and implications of both a political and normative character. An argument based on the findings of the limited use of rights in the SDG reports has political implications in arguing for the necessity to use rights language more explicitly if the goal is to enhance respect for human rights. However, the critical contribution this chapter offers simultaneously speaks to the need to do this with caution, as a simple branding or labelling exercise risks being insufficient, instead leading to window-dressing and empty rhetoric.

The main argument made throughout this chapter is that the discourses on human rights and SDGs do not fit easily together simply as they are, but require analytical work. One of the reasons for this is the tension between a political compromise (the SDGs) and a normative understanding of what legitimate claims of rights-holders entail (human and Indigenous rights). Another key argument is that the misfit between the SDGs and human rights is stressed in double terms when it comes to Indigenous peoples' rights, as there is an inherent tension faced by Indigenous peoples when "engaging and seeking to draw benefit from an international institutional order founded on, and continually reinscribing, their exclusion" (Celermajer & Lefebvre, 2020, p. 13). Adding to that, there are worrying tendencies in work labelled as sustainable development, which simply re-labels colonial practices under the auspices of the so-called just green transition (Ohlsson & Lind, 2025).

However, despite the valid critique of both the SDG and the human rights discourses, I do not see an abandonment of the project of human rights nor the SDGs as the path forward. Politically, more awareness and willingness are needed to take the critique raised here and elsewhere seriously. The solution is not to label all sustainability work as human rights work, which would not solve the problem of window-dressing but rather lead to rights-washing, but, as I have argued throughout this chapter, the language and terminology used are of importance. By not talking about human rights, and, when needed, adding the explicit language of Indigenous rights, the risk of the SDGs' potential to actually lead towards more inclusive and sustainable societies gets lost.

Competing Interests The author has no conflicts of interest to declare that are relevant to the content of this chapter.

References

Agyeman, J. (2013). *Introducing just sustainabilities: Policy, planning, and practice.* Zed Books.

Agyeman, J., Bullard, R. D., & Evans, B. (2002). Exploring the nexus: Bringing together sustainability, environmental justice and equity. *Space and polity, 6*(1), 77–90.

Alston, P. (2005). Ships passing in the night: The current state of the human rights and development debate seen through the lens of the millennium development goals. *Human Rights Quarterly, 27*(3), 755–829.

Arora-Jonsson, S. (2023). The sustainable development goals: A universalist promise for the future. *Futures, 146*, 103087. ISSN 0016-3287

Bentham, J. (1843). Anarchical fallacies. In J. Bowring (Ed.), *The works of Jeremy Bentham* (Vol. 2). William Tait.

Breidlid, A., & Krøvel, R. (Eds.). (2020). *Indigenous knowledges and the sustainable development agenda.* Routledge.

Cambou, D. (2019). The UNDRIP and the legal significance of the right of indigenous peoples to self-determination: A human rights approach with a multidimensional perspective. *The International Journal of Human Rights, 23*(1–2), 34–50.

Celermajer, D., & Lefebvre, A. (Eds.). (2020). *The subject of human rights* (Stanford Studies In Human Rights). Stanford University Press.

Cole, W. M. (2015). Mind the gap: State capacity and the implementation of human rights treaties. *International Organization, 69*(2), 405–441.

Corntassel, J. (2008). Toward sustainable self-determination: Rethinking the contemporary indigenous-rights discourse. *Alternatives, 33*(1), 105–132.

de Jong, S. (2022). Writing rights: Suturing Spivak's postcolonial and de Sousa Santos' decolonial thought. *Postcolonial Studies, 25*(1), 89–107.

Esaiasson, P., Gilljam, M., Oscarsson, H., & Wängnerud, L. (eds.). (2012). Metodpraktikan: konsten att studera samhälle, individ och marknad (4th ed.). Norstedts juridik.

Esquivel, V. (2016). Power and the sustainable development goals: A feminist analysis. *Gender and Development, 24*, 9–23.

Feiring, B., & Hassler, A. (2016). Human rights in follow-up and review of the 2030 agenda for sustainable development. The Danish Institute for Human Rights. ISBN: 978-87-93241-69-5.

Gómez Sánchez, D. (2020). Transforming human rights through Decolonial lens. *The Age of Human Rights Journal, 15*, 276–303.

Grenholm, C.-H. (2007). Introduction. In C.-H. Grenholm & N. Kamergrauzis (Eds.), *Sustainable development and global ethics*. Acta Universitatis Upsaliensis.

Grovogui, S. N. (2006). Mind, Body, and Gut! Elements of a postcolonial human rights discourse. In B. G. Jones (Ed.), *Decolonizing international relations*. Rowman & Littlefield.

Hannum, H. (1995). The status of the universal declaration of human rights in national and international law. *Georgia Journal of International and Comparative Law, 25*, 287–398.

Herlin-Karnell, E. (2023). The constitutional concepts of sustainability and dignity. *Jus Cogens, 5*, 125–148.

Hohfeld, W. N. (1964). *Fundamental legal conceptions as applied in judicial decisions*. Orig. 1919. Yale University press.

Kaltenborn, M., Krajewski, M., & Kuhn, H. (Eds.). (2020). *Sustainable development goals and human rights*. Springer Nature.

McNeil, D. (2007). The ethics of sustainable development. In C.-H. Grenholm & N. Kamergrauzis (Eds.), *Sustainable development and global ethics*. Acta Universitatis Upsaliensis.

Mignolo, W. D., & Walsh, C. E. (2018). *On Decoloniality: Concepts, analytics, Praxis*. Duke University Press.

Muñoz Carmona, L. (2024). Synergies or silos? Exploring human rights considerations in sustainability reporting practices in the Nordics. *Netherlands Quarterly of Human Rights, 42*(1), 71–89.

Mutua, M. (2002). *Human rights: A political and cultural critique*. University of Pennsylvania Press.

Namli, E. (2020). Mänskliga rättigheter när rasismen normaliseras. In P. Bremdal & J. Ohlsson (Eds.), *Mänskliga rättigheter i teori och praktik: från idé till förvaltning* (pp. 71–96). Studentlitteratur.

Ohlsson, J., & Lind, S. (2025). Justice, resilience and indigenous land dispossession in the "Green" transition: Perspectives from Forest Sámi communities in Sweden. In T. Skillington & A. Setti (Eds.), *Climate change resilience across societal contexts: Challenges, Potentials and Visions for the Future*. Palgrave Macmillan.

Österdahl, I. (2023). Agenda 2030 as a source of law: Different shades of soft. In M. Dahlberg, T. Fridström Montoya, M. Hansson, & C. Zetterberg (Eds.), *De lege 2022:* Hållbarhet ur ett rättsligt perspektiv, Iustus förlag, pp. 355–384.

Pathak Broome, N., Bajpai, S., & Shende, M. (2020). Forest rights act, local collectivisation and transformation in Korchi. In A. Breidlid & R. Krøvel (Eds.), *Indigenous knowledges and the sustainable development agenda*. Routledge.

Pogge, T., & Sengupta, M. (2016). Assessing the sustainable development goals from a human rights perspective. *Journal of International and Comparative Social Policy, 32*(2), 83–97.

Purvis, B., Mao, Y., & Robinson, D. (2019). Three pillars of sustainability: In search of conceptual origins. *Sustainability Science, 14*, 681–695.

Shaw, M. N. (2021). International law (9th ed.). Cambridge University Press.

Struckmann, C. (2018). A postcolonial feminist critique of the 2030 agenda for sustainable development: A south African application. *Agenda, 32*(1), 12–24.

Sultana, F. (2018). An(other) geographical critique of development and SDGs. *Dialogues in Human Geography, 8*(2), 186–190.

Toivanen, R., & Cambou, D. (2021). Human rights. In C. P. Krieg & R. Toivanen (Eds.), *Situating sustainability: A handbook of concepts and contexts* (pp. 51–62). Helsinki University Press.

Toivanen, R., Vladimirova, V., & Ojala, C.-G. (Eds.). (2026). *Decolonizing the Sustainable Development Goals*. Palgrave Macmillan.

United Nations Department of Economic and Social Affairs, UN DESA. (2007). United Nations declaration on the rights of indigenous peoples. UN DESA. Retrieved January 15, 2025, from https://social.desa.un.org/issues/indigenous-peoples/united-nations-declaration-on-the-rights-of-indigenous-peoples

United Nations Department of Economic and Social Affairs, UN DESA. (2023). *The sustainable development goals report 2023: Special edition - July 2023*. UN DESA. Retrieved January 15, 2025, from https://unstats.un.org/sdgs/report/2023/

United Nations General Assembly, UNGA. (2015, October 21). *Transforming our world: The 2030 Agenda for Sustainable Development*, A/RES/70/1. Retrieved February 13, 2024, from https://www.refworld.org/legal/resolution/unga/2015/en/111816

United Nations Office of the High Commissioner for Human Rights, UN OHCHR. (2024). *OHCHR and the 2030 Agenda for Sustainable Development*. Retrieved January, 19, 2024.

United Nations Sustainable Development Summit. (2023, August 8). *Fast Facts—What is Sustainable Development?* Blogpost. Retrieved January 15, 2025, from https://www.un.org/sustainabledevelopment/blog/2023/08/what-is-sustainable-development/

Williams, C., & Blaiklock, A. (2016). Human rights discourse in the sustainable development agenda avoids obligations and entitlements: Comment on "rights language in the sustainable development agenda: Has right to health discourse and norms shaped health goals?". *International Journal of Health Policy and Management, 5*(6), 387–390.

Winkler, I. T., & Satterthwaite, M. L. (2017). Leaving no one behind? Persistent inequalities in the SDGs. *The International Journal of Human Rights, 21*(8), 1073–1097.

Winkler, I. T., & Williams, C. (2017). The sustainable development goals and human rights: A critical early review. *The International Journal of Human Rights, 21*(8), 1023–1028.

Woods, K. (2010). *Human rights and environmental sustainability*. Edward Elgar Publishing.

World Commission on Environment and Development (WCED). (1987). *Our Common Future*. Oxford University Press.

CHAPTER 7

Back to the Land of the Future: (Re)Producing a Northern Industrial Heartland and Indigenous Homeland for Sweden's 'Green Transition'

Elisa Maria López

Introduction

In February 2023, a traveler passing through the Tekniska Högskolan station in Stockholm would have encountered the following sight: the long automatic escalator wells, which transport passengers to and from the subway platform eighteen meters underground, entirely wrapped in large full-color advertisements (Fig. 7.1) for the Swedish state-owned mining company LKAB (Luossavaara-Kiirunavaara Aktiebolag). One ad featured a photo of a young woman in mineworker's uniform in an underground tunnel, framed by the slogan "*Shape the future - The greatest challenge – The greatest opportunity – Go North.*" Inside the station, the ads continued with another photo of a woman in hiking gear standing on the summit of a snowy mountain, this one with the slogan "*Our greatest challenge. Your greatest opportunity. 1365 meters below ground, a quest towards a bright and sustainable future has begun*".

The ads were part of LKAB's latest effort to recruit graduates from the institution adjacent to the subway station, Kungliga Tekniska Högskola (KTH Royal Institute of Technology), the nation's largest and most important educational institution for training future civil engineers, architects, urban planners, chemists, physicists, and other technical and scientific experts. As an anthropologist and researcher employed at KTH School of Architecture myself, I encountered the advertisements shortly after returning to Stockholm from a month-long research stay in the "North" depicted in the ad images and

E. M. López (✉)
School of Architecture, KTH Royal Institute of Technology, Stockholm, Sweden
e-mail: elopez@kth.se

R. Toivanen et al. (eds.), *Decolonizing the Sustainable Development Goals*,
https://doi.org/10.1007/978-3-032-14263-4_7

Fig. 7.1 LKAB advertisements in the Tekniska Högskola subway station, Stockholm, February 2023. (Photo: Elisa Maria López)

slogans: specifically, from Kiruna (in North Sámi, *Giron*), Sweden's most northern city in the Sámi Indigenous homeland, Sápmi. Since the 1890s Kiruna has also been the site of the world's largest underground iron ore mine, today operated by the Swedish state-owned mining company LKAB. I have studied the social and spatial impacts of iron mining in Kiruna since 2012 from a historical and ethnographic perspective, attending primarily to the experiences of Kiruna's Indigenous Sámi community upon whose lands the city, mine, and infrastructures were built (López, 2021a).

Against this backdrop, this chapter investigates the social and spatial production of northern Swedish Sápmi as an extractive resource frontier necessary for materializing state visions of national development. How have state and corporate actors produced northern Sweden and Kiruna as ideal spaces for industrialization and extraction? How is this process unfolding in the context of Sweden's industrial 'green transition' agenda, and what are the implications for Sámi Indigenous livelihoods and rights in specific communities? Based on archival research, anthropological fieldwork, and ethnographic interviews conducted in Kiruna between 2022 and 2023, I demonstrate how socio-spatial imaginaries have been integral to (re)producing Kiruna and northern Swedish

Sápmi as future extractive resource frontiers since the nineteenth century. I argue that because today's 'green transition' agenda to decarbonize Swedish industry also involves the large-scale transformations of northern communities to accommodate new residents/industrial workers—but does not address existing structures of inequality that have been preconditions for resource extraction—Sweden's "green transition" risks not only undermining the purported aims of UN Agenda 2030's Sustainable Development Goals, but also social and spatially reinforcing Sámi marginalization, dispossession, and colonial violence.

My argument proceeds across three sections. In the first, I analyze Swedish 'green transition' development as a form of capitalist space production, tracing how long-standing national political and industrial economic goals have produced northern Sweden as an abstract, empty, and valuable space ideal for industrial resource extraction (Mason, 2022). In the second section, I turn to Kiruna's founding as a settler city in 1900, tracing how earlier forms of 'greening' worked to displace and replace native people and nature. In the third section, I examine how state and corporate discourses of contemporary 'green transitions' work to establish continuities with earlier eras of Swedish welfare politics and ideology, noting how the production of northern extractive "welfare frontiers" (Hastrup & Lien, 2020) also marginalized Sámi collective rights. I further trace how state and corporate visions for northern 'green transitions' reinforce Sámi dispossession in places where planned 'green' projects primarily entail the expansion of mining operations.

Back to the Land of the Future

The social, conceptual, and material production of northern Sweden/Sápmi (Lefebvre, 1991; Smith, 2008) as a space of "the greatest opportunity" has a much longer history, dating as far back as the sixteenth century (Fur, 2006; Ojala & Nordin, 2015). Already by 1630, Sápmi's promise and potential as a resource frontier was articulated by crown authorities through analogies with other colonial frontiers like a "Swedish America," or "West Indies of the Swedes," and parallels often made between Sámi and Native Americans (Bäärnhielm, 1976; Ojala & Nordin, 2015, p. 11). At this time, Sápmi's importance as a crown asset was primarily through administrative infrastructures—for example, the Swedish church—and the extraction of taxes from both Sámi and other northern residents. Indeed, it would not be until the nineteenth century that technological advancements in the production of steel made large-scale extraction in far northern Sápmi—such as present-day Kiruna—possible for the first time. As in earlier eras, these technological and infrastructural advancements to extract value from Sápmi entailed a co-production of physical, imaginary (or mental), and social spaces (Lefebvre, 1991, pp. 11–12), the three components of what sociologist Henri Lefebvre called "logico-epistemological" space, "…the space of social practice, the space occupied by sensory phenomena, including products of the imagination such as projects and projections, symbols and utopias" (ibid.). Encountering the subway advertisements, I noted the return of a well-known nineteenth century spatial imaginary of the north—*Framtidslandet*, "The Land of the Future"—that, as I would soon learn, was coming back into fashion through discourses of Sweden's "green transition."

The Stockholm subway advertisements were in fact linked to an upcoming job fair at the KTH Student Union, where LKAB was one among many companies and agencies with on-site exhibition booths. At the fair, LKAB company representatives chatted with students, offering Swedish and English-language company brochures, keychains, and other promotional items. Among these, a magazine titled *Kultje* (LKAB, 2022) caught my eye. The illustrated cover depicted 'the world' from the perspective of northern Sweden: the most prominent places named are Skellefteå, Boden, Piteå, Luleå, Gällivare, and Kiruna—the northern cities most associated with the "green transition" —and the landscape is depicted as one of wind turbines, construction cranes, new factories, LKAB's mines, and the European rocket range (Esrange). To the south, iconic buildings mark the city of Stockholm (Globen Arena) and the Skåne region (the Turning Torso, Malmö). Across from the Nordics and a green rectangle labeled "Europe," small hill-like forms labeled "Asia," "Australia," "Africa," and "South America" dot the horizon like islands in a distant archipelago.

Referencing a more recent techno-spatial frontier, the magazine's introductory essay, "The Silicon Valley of Tomorrow," explains that *Kultje*'s focus is on "climate change, industrial development, and a deeper look at the northernmost parts of Sweden." We read that the word *Kultje* "simply means heavy winds, or gale force winds. And that perfectly describes the situation in Sweden right now." This situation is described as follows:

> Investment plans for the coming decades are set for around SEK 1 trillion of industrial development focused on the northernmost parts of Sweden. And if you listen to those beating the largest drum, this means upwards of 100,000 new residents in northern Norrland.... This is the story about how [sic] Sweden's center is headed north and turning the map of Sweden upside down. (ibid., 2)

That LKAB is one such actor "beating the largest drum" about "green" development is clear in the next article, "Transformation in the North: Back to the Land of Tomorrow" (LKAB, 2022, p. 4). The article's title references a popular nineteenth-century slogan, "The Land of the Future" (*Framtidslandet*) and even cites historian Sverker Sörlin's 1988 dissertation—also titled *Framtidslandet*—on the subject (Sörlin, 1988 [2023]). Describing Sörlin as "one of the voices urging today's optimists to think long term and become more concerned about sustainable social development" (ibid), the article seeks to link frontierism with the "green transition," offering "people living in the Swedish Arctic [a] second chance at building a sustainable land of tomorrow" (LKAB, 2022, p. 4).

While LKAB's article embraces this nineteenth-century frontierism, Sörlin's 1988 dissertation is a critical examination of how geographical imaginaries, agendas, and debates on northern people and environments developed during this period of industrialization. Sörlin characterizes the scientific and political-economic view of northern Sweden of this era as 'geodeterminism,' in which large-scale exploitation of abundant and "cheap" natural resources and energy were viewed as advantages in the competition between industrialized nations, paving the way for natural resources to become part of a Swedish "nationalist credo" (Sörlin, 1988 [2023], p. 197). Sörlin also underscores the tight links between scientific research and economic interests in the north throughout history, both before and during the nineteenth century, noting significant links between large-scale resource extraction and its justification by political and industrial elites through views of northern people as underdeveloped and "uncivilized," advanced through pseudoscientific racial biology and views of industry as a higher, "civilizing" force (ibid., 261).

These critiques of "The Land of the Future" ideology, however, are nowhere to be found in LKAB's 2022 article. Indeed, LKAB and its representatives are key actors identified in Sörlin's text as advocates of late nineteenth- and early twentieth-century industrial geodeterminism, racial biology, and "internal colonization" of northern environments and people (ibid., 283). In celebrating the frontier imaginary of the "The Land of the Future," LKAB not only consciously ignores decades of social, scientific, and political critique of elite views of northern Sweden/Sápmi, but openly advocates for a return to frontierism as a way of remaking space and time, a "travelling theory" that reaches "backward as well as forward in time energizing old fantasies, even as they embody their impossibilities" (Tsing, 2003, pp. 5100–5101). In the following section, I too look back at the nineteenth and early twentieth centuries, specifically, to LKAB's founding of the city of Kiruna and establishment of the mine as the production of a settler colonial frontier, in which the erasure of both native people and ecology was a prerequisite to building a "Land of the Future."

Greening the Settler City

Kiruna is a twentieth-century industrial New Town, or "model city" (*modellstad*), founded in 1900 by the mining company LKAB. The city's early architectural and urban history is popularly credited to its 'founding father' Hjalmar Lundbohm, the first mine director of LKAB Kiruna, who was famously inspired by industrial company towns he had visited in the United States and England. Kiruna's first city plan, designed by planner Per O. Hallman, was one of the first applications of Garden City principles—adapting an urban plan to the natural terrain—in Swedish urban planning. Additionally, Lundbohm commissioned architect Gustaf Wickman to design homes for workers near his own in the Company Neighbourhood (*Bolagsområdet*), a residential area reserved exclusively for LKAB administrators, employees, and their families.

Kiruna's urban history also reveals how nineteenth-century ideas of northern nature and culture co-produced social and physical spaces for southern settlers to feel at home in the homelands of the areas' Indigenous Sámi and Tornedalian peoples. One of the ways this settler domesticity was produced was through the creation of a new nature, or what anthropologist Rubah Salih and political scientist Olaf Corry, writing in the context of Palestine, term "settler-Nature" (Salih & Corry, 2022). As settler colonialism seeks to displace and permanently replace a land's native inhabitants, the erasure of both native people and environments is a prerequisite to establishing settler societies (Wolfe, 2006). Salih and Corry argue that settler-Nature is a further product of the 'canonical' settler colonial process: first, elimination of native people and ecological habitats, followed by a discursive and legal production of the environment as a 'wasteland,' and ultimately the 'redemption' of the land as 'settler-Nature' through settler technologies, ethics, and labor (ibid., p. 389).

Settler colonialism in Kiruna followed a similar pattern. As frontier imaginaries of northern Sweden gained popularity, in 1886 all traditional Sámi land in the north above the "cultivation border" —including Kiruna—was declared crown property (Össbo & Lantto, 2011). At this time, the state also took over administration of reindeer herding, leading to the expansion of a colonial administrative system which would exert forms of biopolitical control over reindeer herding Sámi from 1885 until 1971 (Lantto, 2012). For Lundbohm and other early twentieth-century majority elites concerned with the so-called Sámi question, colonial views of Sámi people and culture as a lesser form of civilization were enforced, among other ways, through ideals of Sámi 'authenticity.' In the eyes of colonial elites, the true 'authentic' Sámi livelihood was reindeer herding, and 'authentic' reindeer herders those who lived and worked in 'nature,' separate from industrial or agricultural society. Yet this did not reflect the diversity of Sámi culture and livelihoods in northern Sweden, for example Sámi who practiced other livelihoods or a mix of livelihoods. Through the legal re-definition of Sámihood and who was allowed to keep reindeer in the 1928 Reindeer Herding Act (Össbo & Lantto, 2011), many Sámi lost the 'right' to herd reindeer, and in turn were dispossessed of land rights in their traditional home areas. Those who kept the 'right' to herd, however, became subject to even greater colonial control. Many families were forcibly resettled after the border closures between Sweden and neighboring Norway and Finland, exacerbating assimilation and, in turn, freeing up land for settlers (Lantto, 2010; Labba, 2020).

Urban space and modern buildings were also considered threats to Sámi 'authenticity' by colonial elites like Lundbohm, who believed Sámi would abandon the reindeer herding livelihood if given the opportunity to reside or spend extensive amounts of time in modern homes (López, 2021a, pp. 128–29). Accordingly, in Kiruna's early years, reindeer herding Sámi were not allowed to live in modern houses, in the city, or stay in the city's hotels, despite their active social and business connections with urban residents. Displaced and excluded from now urbanized homelands, Kiruna's Sámi community nevertheless persisted, drawing on kin, friendship, and other social relations with urban residents to claim space for themselves in the settler city, and eventually establishing the first hotel and community center for Sámi in the city in the 1960s (ibid., pp. 138–39).

Sámi dispossession and displacement were also accompanied by the creation of a new settler-Nature in Kiruna, which found ample expression in this industrial "Garden City in the Lappmarks" (Bedoire, 1974, p. 144). As the main actor tasked with planning and developing urban Kiruna, LKAB further contributed to creating 'settler-Nature' in Kiruna through urban green spaces and gardens, in keeping with the "Garden City" ideal of urban "greening" as means to improve society (Angelo, 2021). Until the 1960s, for example, LKAB maintained both company greenhouses and kitchen gardens, tended to by a company master gardener and staff. Here, company gardeners cultivated a diversity of edible plants, including grapes, which were made into wine and served to visiting businessmen at the company hotel or sometimes given in bunches to employees with their Christmas bonuses as gifts (Fig. 7.2). Local boys were also paid by LKAB to maintain green spaces in the Company Neighbourhood during summers, landscaped with grasses and flowers grown in the company's greenhouses for this purpose (Fig. 7.3).

Private gardens were also sites where settler-Nature and domesticity were cultivated. Most city residents maintained garden plots with potatoes, radishes, spinach, beets, and flowers in Kiruna's early years (Simmons, 1910), and until the 1960s, LKAB employees could find gardening tips from the company's master gardener in issues of *Skip*, LKAB's employee magazine. In 2013, an environmental report commissioned by LKAB ahead of the relocation of Lundbohm's private home (SWECO Environment, 2013) also found several non-native plants growing in Lundbohm's home garden, including roses, black currant, and North American blue spruce. Upon request, I was shown a letter to Lundbohm in LKAB's archives explaining the origins of five Swiss Pine (*Pinus cembra*) trees and a European Larch (*Larix Europa*, which accompanied a package of seeds sent by Director Erik Ljungberg of the Stora Kopparberg copper mine in Falun:

> When I was down in St. Moritz last summer I discovered, above the timberline 6,600 feet above sea level two tree species growing a further 600 feet higher up, which had surprising branch structure and growing power. In special packages I send you seeds from these trees, and I hope that they will, when you give them your interest and care, spread joy to your descendants in Kiruna.
>
> (Letter to Hjalmar Lundbohm from Erik Ljungberg, April 19, 1904)

Notably, in this seed exchange between the two mine directors, nature is not treated as a site of extraction or production—the mine directors' daily business—but rather as space of leisure, scientific discovery, and paternal domesticity. As ethnologists Jonas Frykman and Orvar Löfgren (1987) observe, at the time the letter was written, Swedish cultural views of nature as primarily a

Fig. 7.2 Issue of *Kultje* magazine and other promotional items from LKAB's booth at the KTH Student Union, February 2023. (Photo: Elisa Maria López)

sphere of production and labor were being replaced by upper- and middle-class views of 'nature' at this time, becoming an external sphere of capitalist leisure and non-work, a counter-space to the domestic and labor spheres where culture and modernity were cultivated (ibid., pp. 68–69). The socio-spatial ideals were not only incorporated as "building blocks" (ibid.) of Sweden, but also found expression in modern city planning models which sought to unite nature and culture, or "town and country" (Porfyriou, 1992), as an urban design aesthetic, and which influenced many planners across Europe and the Nordics, including Per O. Hallman.

Indeed, as a renowned "cultural patron" (*kulturmecenat*) of the arts and sciences (Persson, 2018), Hjalmar Lundbohm not only cultivated exotic plants in his home garden, but also funded scientific research regarding Kiruna's ecology on a greater scale. In 1908, Lundbohm hired botanist Herman G. Simmons on behalf of LKAB, to study how settlers were changing Kiruna's plant life (Simmons, 1910). The foreword to Simmons' *The Flora and Vegetation of Kiruna: A Phytogeographical Study with Special Regard to the Influence of*

Fig. 7.3 Wine grapes being cultivated in an LKAB greenhouse, Gällivare Municipality, ca. 1917–1972. (Photographer: Lea Wikström. Source: Norrbottens museum)

Culture is also written by Lundbohm, and where he explains why he saw this research as necessary:

> Kiirunavaara and Luossavaara lie in a wasteland (*ödemark*)...It was known that a large population would soon gather here, and it was important that they live under reasonably good conditions. The harsh climate, the great distance from the country's central cultural areas, the isolation from socializing with people other than those who in one way or another deal with the processing of the ore, and the resulting monotony, the long and gloomy polar night – all this was designed to make the place less attractive as a place of residence for people with conventional interests and demands on life, and one could fear that it would be difficult for the population to find the facilities that everyone has the right to demand, and which is a necessary condition for good work to be performed.

Therefore, it was also important to carefully study the possibilities for creating a reasonably satisfactory environment for the people who will choose Kiruna as a place of residence. (Lundbohm in Simmons, 1910, pp. 5–6)

In this text, Lundbohm articulates views of the period, as observed by Sörlin (1988 [2023], p. 261), which strongly link industrialization with "culture," Kiruna's environment is considered a 'wasteland,' amenable for technoscientific transformation into "a reasonably satisfactory environment" for future settlers (Simmons, 1910, p. 6). Although Lundbohm was keenly aware of the impact settlers were having on the land and the fragility of the Arctic environment, he was primarily concerned with sustaining the new settler community and modifying the environment toward this end. A further purpose of Simmons' study was thus to explore possibilities for introducing new species to replace native ones. Notably, these new species were not only for practical use (food, lumber), but also for improving the landscape aesthetically, and Lundbohm hoped the study would identify "which types of grasses were best suited for lawns and fodder, [and] which ornamental plants one should cultivate" (ibid.). Emphasizing the need to plant lawns and cultivate ornamental flowers, Kiruna's native ecology—an unappealing 'wasteland'—could be redeemed through ecological and aesthetic "greening," aided by scientific knowledge and technologies like greenhouses. In doing so, LKAB sought to transform Kiruna's environment to provide settlers with the "facilities that everyone has the right to demand, and which is a necessary condition for good work to be performed" (ibid.). Yet as noted, this 'right' to an ideal environment did not apply to 'everyone,' particularly Sámi displaced from their homelands, excluded from settler city, or stripped of their traditional livelihoods.

This social and environmental history does not only remain relevant for understanding how late nineteenth- and early twentieth-century industrial settler colonialism unfolded in Kiruna and across Sápmi. As I show in the following section, it is also relevant for understanding the persistent power of the northern resource frontier imaginary within modern political and development agendas—from the Social Democratic welfare state to today's 'green transition'—justifying new forms of northern exploitation and transformation for the sake of 'the greater good.'

Making a Northern Welfare Frontier

As Sörlin observed in his 1988 study, there was also a relationship between the north's ideological production as resource frontier and the project of Swedish modernization, specifically, the construction of the twentieth-century Social Democratic welfare state, a utopian "land of the future" in its own right (Sörlin, [1988] 2023, p. 21). Although Sweden has steadily become more neoliberal since the 1980s, legacies of Social Democratic welfare governance and ideology remain cornerstones of modern Swedish society and national identity. Accordingly, long-standing associations between industry and welfare also remain significant, as *Kultje* made clear. The magazine's concluding essay, "The Future is in Norrbotten" (LKAB, 2022, p. 91), is penned by Göran Persson, former Social Democratic Prime Minister of Sweden (1996–2006), and from 2017 to 2023, chairman of the board of LKAB. In the text, Persson explains that to achieve decarbonization, we must not only mine more, but build more in the North:

> Over the next few decades, LKAB will be making unimaginable investments in our communities. We will be able to recoup all these investments, both from a financial and environmental perspective…We will create new facilities and new jobs. We will build new cities. We will profit and we will contribute to the Swedish treasury. All whilst contributing to a better world…Welcome to Norrbotten. There are jobs here. The hardest, most challenging jobs are here. The future is here. (LKAB, 2022, p. 91)

To a Swedish audience, Persson's message would have a familiar ring to it: as Prime Minister from 1996 to 2006 and leader of the Social Democratic party until 2007, Persson was a powerful figurehead and advocate of the Social Democrats' "green welfare state" agenda, transforming Sweden into a "modern environmental" nation (Hildingsson, 2014). In the 1960s, Sweden became one of the first countries in the world to adopt "ecological modernization" as a socioeconomic and political agenda, which focused on enacting environmental imperatives across all scales and sectors—from developing energy-efficient technologies to strengthening legal frameworks for environmental conservation (ibid.). It is within this context that today's notion of a system-wide "green transition" emerged: the reduction and elimination of fossil fuel use through economic, political, institutional, and social change (Hildingsson et al., 2019, p. 912).

With an established track record of "translating" natural resources into national welfare and "translating dreams of nation-building into promises" (Sejersen, 2020, p. 4) as Prime Minister, Persson is also a significant "broker of hope" (ibid.) in his role as Chairman of the Board of LKAB. In this broker role, Persson can "translate" extraction into sustainability; decarbonization becomes not only compatible with extraction, but a shorthand for teleological economic growth, development, and welfare. Industrial expansion is thus framed as a "win-win" scenario, as these "unimaginable investments" from industry will not only benefit local communities and the nation, but also the world. In doing so, Persson links today's "green transitions" and earlier eras of the Social Democratic welfare state where LKAB was a major industrial community builder (Isacson, 2007).

Like today's "green transitions," the agenda of the Social Democratic People's Home (*Folkhemmet*, ca.1937–1976) was a "techno-optimistic" one (ibid., p. 43), characterized by the high modernist belief that scientific rationalization and organization of space and society would generate a utopian, democratic, and egalitarian society. With the state harnessing the forces of capitalist economic growth and consumption for "the benefit of all" (Mattsson & Wallenstein, 2010), social reformers focused on large-scale modernization and rationalization of Sweden's built environment—from new town planning to innovative domestic architecture—as means to quite literally build a better society. Notable efforts included the construction of one million housing units between 1965 and 1975, the so-called Million Program (*Miljonprogrammet*), to accommodate a growing urban workforce increasingly drawn from the rural countryside. Simultaneously, industrial companies like LKAB were given great freedom to build and develop urban areas for their work forces, and were seen as key actors for urbanizing the Swedish countryside at this time (Curman, 1944; Isacson, 2007, p. 215).

Today's state and industry actors have also articulated visions of northern space and society for the "green transition" linked to welfare and community development. Faintly echoing Lundbohm's words from 1910, a 2022 report by the Ministry of Climate and Enterprise describes the problem of housing shortages and an estimated 100,000 new northern residents/workers required within the next fifteen years as follows:

> Housing issues can be viewed in two ways. An increased volume is central to cope with increased occupancy. At the same time, well-executed community building with attractive housing is part of creating increased attractiveness so that more people want to come to a new place of interest. (Näringsdepartamentet, 2022, p. 23)

Toward this end, in 2021 the Council For Sustainable Cities initiated the *Visioner: I Norr* (Imagining: The North) program, organized by the museum and government agency ArkDes, the Swedish Centre for Architecture and Design. *Visioner: I Norr* matches northern municipalities with multidisciplinary teams of architects, designers, and other actors to design more attractive cities. Proposals were solicited through two open calls, *Visioner: I Norr I* in 2021 and *II* in May 2024. The 2024 competition call for *Visioner: I Norr II* read:

> In Norrbotten and Västerbotten, some of the biggest investments ever in green transitions, technology and infrastructure are currently taking place. The investments place great demands on more housing, attractive living environments and long-term sustainability. How can we create sustainable places and housing in a short time, while at the same time making use of existing resources and creating long-term good living environments with good quality? How can we work and think in new ways?
> (Rådet för hållbara städer, 2024, p. 3)

Although all the cities in focus for both calls are in northern Sweden/Sápmi—Umeå, Skellefteå, Boden, Luleå, Gällivare, and Kiruna—neither brief mentions Sápmi, the Sámi, or reindeer herding, a constitutionally protected national interest. While there is a mention of "complex and conflicting interests" in the second call (Rådet för hållbara städer, 2024, p. 4), the program's focus is to "generate future scenarios and visions that can support municipalities in the development of living environments that are beautiful and resilient, attractive and fossil-free, innovative and inclusive" (Rådet för hållbara städer, 2021, p. 1). Both calls are also associated with the European Union's "New European Bauhaus Initiative," which aims to "connect the European Green Deal to our daily lives and living spaces. It calls on all Europeans to imagine and build together a sustainable and inclusive future that is beautiful for our eyes, minds, and souls" (European Commission, 2022). Yet an integral part of the European Green Deal is the Critical Raw Materials Act, which links European security and the extraction of materials considered "indispensable for the EU economy and a wide set of necessary technologies for strategic sectors such as renewable energy, digital, aerospace and defense" (European Union, 2024). The day in January 2023 I arrived in Kiruna to conduct fieldwork, for example, LKAB announced it had discovered and planned to mine one of Europe's largest deposits of phosphorous, a critical raw material and rare earths element (REE). At the press conference, LKAB CEO Jan Moström and Minister of Energy and Industry Ebba Busch donned miners' uniforms and helmets as they made the announcement in an auditorium 540 meters underground at the LKAB mine. The press conference also coincided with a nearby meeting of the European Council of Ministers occurring in Kiruna the same day. Moström described the Per Geijer deposit as "...good news, not only for LKAB, the region and the Swedish people, but also for Europe and the climate. This is the largest known deposit of rare earth elements in our part of the world, and it could become a significant building block for producing the critical raw materials that are absolutely crucial to enable the green transition. We face a supply problem. Without mines, there can be no electric vehicles" (LKAB, 2023).

For Kiruna's Sámi *čearru* (reindeer herding collectives) Laevas and Gabna, on whose lands the Kiruna mine and city were built—and where the Per Geijer deposit lies—the announcement came as a shock, however. Neither had been notified about LKAB plans to mine Per Geijer until the press conference, following a corporate communication pattern I had observed during my PhD fieldwork a decade earlier: LKAB's announcement of new projects, major company decisions, or environmental impacts with widespread local effects through the media, and without informing key local stakeholders beforehand (López, 2021a, p. 221). Mining Per Geijer would not only destroy migration routes for both *čearru* and split Gabna's herding lands in half, but also compound a large-scale loss of land due to LKAB's "urban transformation" of Kiruna, a mining-induced displacement and resettlement megaproject ongoing since 2004 (López, 2021b). Herders from both Gabna and Laevas I spoke to in the days after the press conference expressed significant anxiety and stress, as some feared the cumulative pressures from industry would lead to the complete collapse of the ecosystem necessary to sustain reindeer. LKAB's announcement was also met with mixed reactions by Kiruna's municipal urban planners and developers. Currently, three thousand fly-in-fly-out (or "car-in-car-out") workers commute to Kiruna per week, living in barracks or in shared single-family homes purchased by companies due to housing shortages. While planners hoped the projected 7000 new workers would move to Kiruna permanently, some were also concerned building under boom conditions might lead to a housing surplus if this new demographic did not materialize. Simultaneously, a thousand jobs remain unfilled in Kiruna Municipality, as the tough housing market and general state of social services, all understaffed and underfunded, have long affected labor migration to the municipality.

While all the cities in the *Visioner* call have—like many communities across Sweden's rural countryside—faced problems with growing outmigration, a dwindling tax base, housing shortages or surpluses, and cuts to welfare, how programs like *Visioner: I Norr* or *New European Bauhaus* will improve local structural concerns remains elusive. Yet as in earlier eras, today's deficient northern regions and cities are re-produced as "coveted places, envisaged by various actors as sites of potential" (Li, 2014, pp. 13–14), to be redeemed primarily through technical and aesthetic interventions. We also find a re-emergence of a frontier imaginary associated with earlier eras of Social Democracy, what anthropologists Frida Hastrup and Marianne Lien (2020) term "welfare frontiers": the transformation of Nordic Arctic regions into spheres of resource extraction in the service of local and national welfare. This coupling of welfare and resource extraction, Hastrup and Lien argue, has produced attitudes toward northern extraction as "benign" (Hastrup & Lien, 2020, p. vi), as the use value and 'availability' of northern land is entangled with a "welfarist, egalitarian morality" and manifested implicitly in existing and proposed extractive projects.

Yet as Sámi political scientist Rauna Kuokkanen observes, notions of individualism and egalitarianism within Nordic welfare ideologies have thus "tended to ignore and curtail the collective dimension of the Sámi as a separate people with distinct rights within the state" (Kuokkanen, 2006, p. 5). For Nordic Sámi, pressure to conform and assimilate can therefore be magnified (ibid.), as Indigenous rights are described as "privileges" and posited as counter to welfare morality of "imagined sameness," or equality as sameness (Gullestad, 2002). In today's era of environmental crisis and "green transition" solutions, Swedish Sámi face substantial challenges in protecting their human rights as they did in the industrial nineteenth century, as some futures continue to be seen as more worthy of sustaining than others, and ecological protection becomes incorporated into existing structures of settler colonialism (Erickson, 2020).

Conclusion

The production of northern Sweden/Sápmi and Kiruna as a promising "green" industrial frontier, as I have shown, is a socio-spatial abstraction with a much longer history. Today, this abstraction both facilitates state and corporate power and articulates visions of a future north to primarily non-local audiences. Focusing on industrial socio-ecological transformations experienced by Kiruna's Sámi community from the late nineteenth century to the present, in this chapter I have demonstrated the relationship between the production of the north as a resource frontier and the material and ideological production of the Swedish nation, as well as the implications of this dynamic for human rights and welfare.

Yet many questions remain unanswered: How do the ambitions of the UN Agenda 2030 Sustainable Development Goals—purported cornerstones of the EU New Green Deal and Sweden's "green transition"—apply to local people and places in the world like Kiruna whose primary 'contribution' to global sustainability are as sites of resource extraction? Given these conditions, how would goals like "Sustainable cities and communities" (SDG 11), "Responsible consumption and production" (SDG 12), "Life on land" (SDG 16), and "Peace, justice, and strong institutions" (SDG 16) be possible in Kiruna? Recent events indicate that wide gaps remain between these goals and their implementation. On January 23, 2025, the Swedish Government invoked Chapter 11, §15 of the Planning and Building Act to force Kiruna Municipality—against its will and legal planning 'monopoly'—to adopt a detailed area plan allowing the Australian mining company Talga to extract graphite, a critical raw material necessary for decarbonizing technologies in steelmaking (electric arc furnaces) and vehicles (electric batteries). In a press release, Minister of Infrastructure and Housing Andreas Carlsson cited the "extraction of valuable substances and materials" as the basis for the decision (Regeringskansliet, 2025). In addition, Minister of Energy Enterprise Ebba Bush—who wore a miner's uniform at LKAB's underground press conference in January 2023—emphasized the importance of the Talga mine for the "green transition":

> Sweden has the minerals needed and are absolutely crucial for digital and green transition. We are known for standing for sustainability and good working conditions, especially in the minerals sector. The graphite that Talga intends to produce is a key material in the manufacture of batteries. (Busch in press release, January 23, 2025)

Busch's representation of Sweden as a role model for sustainability, industry, and worker's rights, however, stands in contrast to the reality that Kiruna has now become what anthropologist Aihwa Ong (2005) calls a zone of 'graduated sovereignty,' a place and population within a state whose governance is dictated by global market forces. Concerningly, the approval of Taiga's mine against the wishes of Kiruna Municipality and its residents indicates Sweden's "green transition" may not only undermine local social and environmental sustainability but also democracy if communities like Kiruna must be constantly re-produced as frontier "spaces of exception" (Rifkin, 2014) where regular forms of state law do not apply. Ultimately, the ambitions and values of national and intergovernmental climate policies risk becoming little more than political slogans of the current era, when proposed solutions to today's climate emergency cannot—or will not—ensure equitable, meaningful, and everyday sustainable futures for marginalized places and peoples.

References

Angelo, H. (2021). *How Green Became Good: Urbanized Nature and the Making of Cities and Citizens*. University of Chicago Press.

Bäärnhielm, G. (1976). *I Norrland hava vi ett Indien: Gruvdrift och kolonisation i Lappmarken under 1600-talet*. Ordfront.

Bedoire, F. (1974). En arkitekt och hans verksamhetsfält kring sekelskiftet: Gustaf Wickmans arbeten 1884–1916. Diss. Stockholm: Stockholms universitet.

Curman, J. (1944). Industriens arbetarebostäder. Stockholm: Industriens utredningsinstitut.

Erickson, B. (2020). Anthropocene futures: Linking colonialism and environmentalism in an age of crisis. *Environment and Planning D: Society and Space, 38*(1), 111–128.

European Commission. (2022, September 7). The New European Bauhaus Initiative. https://build-up.ec.europa.eu/en/resources-and-tools/links/new-european-bauhaus-project

European Union. (2024, April 11). Regulation (EU) 2024/1252 of the European Parliament and of the Council of 11 April 2024 establishing a framework for ensuring a secure and sustainable supply of critical raw materials and amending Regulations (EU) No 168/2013, (EU) 2018/858, (EU) 2018/1724 and (EU) 2019/1020 (Text with EEA relevance). https://eur-lex.europa.eu/eli/reg/2024/1252/oj/eng

Frykman, J., & Löfgren, O. (1987). *Culture builders: A historical anthropology of middle-class life*. Rutgers University Press.

Fur, G. (2006). *Colonialism in the margins: Cultural encounters in new Sweden and Lapland*. Brill.

Gullestad, M. (2002). Invisible fences: Egalitarianism, nationalism and racism. *Journal of the Royal Anthropological Institute, 8*(1), 45–63.

Hastrup, F., & Lien, M. E. (2020). Welfare Frontiers? Resource practices in the Nordic Arctic Anthropocene. *Anthropological Journal of European Cultures, 29*(1), v–xxi.

Hildingsson, R. (2014). Governing decarbonisation: The state and the new politics of climate change. Diss. Lund University.

Hildingsson, R., Kronsell, A., & Khan, J. (2019). The green state and industrial Decarbonisation. *Environmental Politics, 28*(5), 909–928.

Isacson, M. (2007). *Industrisamhället Sverige: Arbete, ideal och kulturarv*. Studentlitteratur.

Kuokkanen, R. (2006). Indigenous peoples on two continents: Self-determination processes in Saami and first nation Societie. *Native American Studies, 20*(2), 1–6.

Labba, E. A. (2020). *Herrarna Satte Oss hit: Om Tvångsförflyttningarna i Sverige*. Norstedts.

Lantto, P. (2010). Borders, citizenship and change: The case of the Sami people, 1751–2008. *Citizenship Studies, 14*(5), 543–556. https://doi.org/10.1080/13621025.2010.506709

Lantto, P. (2012). *Lappväsendet: Tillämpningen av svensk samepolitik 1885–1971*. Centrum för Sámisk forskning, Umeå Universitet.

Lefebvre, H. (1991). *The production of space*. Basil Blackwell.

Li, T. (2014). *Land's end: Capitalist relations on an indigenous frontier*. Duke University Press.

LKAB. (2022). *Kultje* nr.1. Luleå Grafiska.

LKAB. (2023). Europe's largest deposit of rare earth metals located in Kiruna area. Press Release. Retrieved October 21, 2024, from https://lkab.com/en/press/europes-largest-deposit-of-rare-earth-metals-is-located-in-the-kiruna-area/

López, E. M. (2021a). Transforming Kiruna: Producing space, society, and legacies of inequality in the Swedish ore fields. PhD diss., Acta Universitatis Upsaliensis.

López, E. M. (2021b). Stadsomvandlingen i Kiruna: Planering och politik i ett urbant megaprojekt. In G. Körling & S. B. Ullberg (Eds.), *YMER: Megaprojekt—Kritiska perspektiv på storskalig infrastruktur* (pp. 77–95). Svenska Sällskapet för Antropologi och Geografi.

Mason, A. (2022). *Arctic abstractive industry: Assembling the valuable and vulnerable north* (Vol. 5). Berghahn Books.

Mattsson, H., & Wallenstein, S.-O. (2010). *Swedish modernism: Architecture, consumption, and the welfare state*. Black Dog Publishing.

Näringsdepartamentet. (2022). Rapport från samordnaren för samhällsomställning vid större företagsetablering och företagsexpansioner i Norrbotten och Västerbotten. Regeringskansliet.

Ojala, C.-G., & Nordin, J. M. (2015). Mining Sápmi: Colonial histories, Sámi archaeology, and the exploitation of natural resources in northern Sweden. *Arctic Anthropology, 52*(2), 6–21.

Ong, A. (2005). Graduated sovereignty in South-East Asia. In J. X. Inda (Ed.), *Anthropologies of modernity* (pp. 81–104). Blackwell Publishing Ltd.

Össbo, Å., & Lantto, P. (2011). Colonial tutelage and industrial colonialism: Reindeer husbandry and early 20th-century hydroelectric development in Sweden. *Scandinavian Journal of History, 36*(3), 324–348.

Persson, C. (2018). *Hjalmar Lundbohm: Ledare, Samhällsbyggare och Kulturmecenat*. Älven kultur.

Porfyriou, H. (1992). Artistic Urban Design and cultural myths: The Garden City idea in Nordic countries, 1900–1925. *Planning Perspectives, 7*(3), 263–302.

Rådet för Hållbara Städer. (2021). Öppen Utlysning: New European Bauhaus. Visioner: I Norr. ArkDes.

Rådet för Hållbara Städer. (2024).Visioner: Umeå, Skellefteå, Luleå, Boden, Gällivare, Kiruna [Visioner: I Norr II]. ArkDes.

Regeringskansliet. (2025). Pressmeddelande från Landsbygds- och infrastrukturdepartementet: Regeringen förelägger Kiruna kommun att anta detaljplan som möjliggör grafitgruva. https://www.regeringen.se/pressmeddelanden/2025/01/regeringen-forelagger-kiruna-kommun-att-anta-detaljplan-som-mojliggor-grafitgruva/

Rifkin, M. (2014). The frontier as (movable) space of exception. *Settler Colonial Studies, 4*(2), 176–180.

Salih, R., & Corry, O. (2022). Displacing the Anthropocene: Colonisation, extinction and the unruliness of nature in Palestine. *Environment and Planning E: Nature and Space, 5*(1), 381–400.

Sejersen, F. (2020). Brokers of hope: Extractive industries and the dynamics of future-making in post-colonial Greenland. *Polar Record, 56*, 1–11.

Simmons, H. G. (1910). *The Flora and Vegetation of Kiruna: A Phytogeographical study with special regard to the influence of culture [Floran och vegetationen i Kiruna: En växtgeografisk studie med särskild hänsyn till kulturens inflytande]*. Nordiska bokhandeln i distr.

Smith, N. (2008). *Uneven development: Nature, capital and the production of space.* University of Georgia Press.
Sörlin, Sverker. ([1988] 2023). Framtidslandet.: Teg Publishing.
SWECO Environment AB [Werdin, Lisa and Strandell, Gry]. (2013). *Rapport: LKAB Växtinventering Hjalmar Lundbohmsgården.* Kiruna.
Tsing, A. L. (2003). Natural resources and capitalist Frontiers. *Economic and Political Weekly, 38*(48), 5100–5106.
Wolfe, P. (2006). Settler colonialism and the elimination of the native. *Journal of Genocide Research, 8*(4), 387–409.

CHAPTER 8

Indigenous Lessons on How to Harvest a Polar Bear: Exploring Sustainability and Sense-making in a Post-growth Economy

Benedikte Brincker

Introduction

This chapter explores Indigenous sense-making of nature and analyses of how it offers insights into a post-growth economy. In doing this, it employs Karl Weick's theory of sense-making (Weick, 1995) in an Indigenous setting—the Indigenous Inuit population of East Greenland—and takes its methodological inspiration from Barbara Czarniawska's method of shadowing.

While the term 'Indigenous people' covers very diverse groups of people, a common characteristic of Indigenous people is their position of being non-dominant (Mortensen and Barten, 2016). The position of Indigenous people as non-dominant raises the question of foreign rule, which in turn leads into post-colonial studies. In recent years, a post-colonial interrogative has emerged in organisation and management studies. It discusses post-colonialism as embodying three perspectives: an epistemic critique, a historical moment, and a condition experienced within post-colonial times (Jack et al., 2011). The epistemic critique involves the concomitant view that traditional western academia suffers from parochialism when it takes western accounts as accurate descriptions of non-western contexts (Xu, 2008).

This chapter has previously appeared in a slightly different version in the *Arctic Yearbook* 2024.

B. Brincker (✉)
Department of Sociology, University of Copenhagen, Copenhagen, Denmark
e-mail: bebr@soc.ku.dk

R. Toivanen et al. (eds.), *Decolonizing the Sustainable Development Goals*,
https://doi.org/10.1007/978-3-032-14263-4_8

The research that informs this chapter addresses this critique methodologically by drawing on the works of Barbara Czarniawska and theoretically by taking its inspiration from the sociologist Karl Weick. In line with Czarniawska, it follows the action instead of the actor, thus advocating an action-focused and process-oriented approach (Czarniawska, 2004; Weick, 1995; Weick et al., 2005). Against this background, the chapter engages with the hunting and sharing of the catch of polar bears in East Greenland and analyses it using Karl Weick's theory of sense-making. The chapter employs the term 'harvesting' to capture the cultural, social, and economic aspects of how to share the catch, i.e. the polar bear. It does so with the aim of exploring how Indigenous lessons on harvesting a polar bear offer insight into sense-making in a post-growth economy.

The chapter focuses on polar bears as opposed to other forms of catch because it allows us to engage with the enactment of this particular animal as prey, national symbol of Greenland, and as a symbol of environmental degradation. Each enactment is part of a cognition that involves sustainability. However, in the context of Indigenous practices on how to hunt and harvest a polar bear, it would be a mistake to conceptualise sustainability as the polar bear's sustainability as a species. Among the Indigenous Inuit population in East Greenland, sustainability pertains to humans. It pertains to the Indigenous people living there. In recent decades, this group has seen the effects of climate change first hand. They have seen bears that earlier would not come near towns or villages now wander through these places in search of food as they have been deprived of their prey such as seals. Illorqortormiut, the northernmost town in East Greenland, has been especially bothered by frequent appearances of groups of polar bears.

Against the backdrop of climate change, in recent decades alternatives to growth have received increasing scholarly attention. These alternatives cover but are not limited to the notions of green growth, de-growth, and post-growth. They have in common a critical perspective on economic growth linking it to environmental degradation under capitalism. However, they differ in other aspects. Green growth seeks to decouple environmental degradation from economic growth by improving the efficiency of production through technological and market innovations (Sandberg et al. 2019; Bauhardt, 2014). So green growth offers a catalogue of potential improvements to capitalism rather than a critique of it. In contrast, de-growth calls for a substantial reduction in growth especially in the western part of the world, while post-growth invites us to reflect on the growth imperative completely.

These debates open up to critique that is not limited to capitalism, production and consumption under capitalism, or the organisation of capitalist society (Baker, 2019; Chandler, 2018). In this perspective, it makes little difference what are heralded as the causal drivers of environmental degradation. Regardless of where scholars position themselves and whether they highlight capitalism, colonialism, imperialism, mass consumerism, modernity, or enlightenment, as primary causes, the conclusion is common across them (Chandler, 2019;

Bonneuil & Fressoz, 2017). There is no longer a separation between humans, nature, and culture (Chandler, 2019; Baker, 2019; Colebrooke, 2014).

The present chapter engages with growth and a critique thereof from a profoundly empirical point of departure. Following Czarniawska's method of shadowing and her engagement with the work of Weick, this is an invitation to reflect 'on your feet' (Czarniawska, 2005). This implies that one needs to engage with Indigenous people and their practices, not as examples of alternative futures, but as a way of life that exists. While this approach sees Indigenous peoples as embedded in nature, it does not in any way include portraying Indigenous people as having a unique bond to nature or being guardians of nature. The latter constitutes an approach that may easily involve the kind of romanticism that tends to prevail in colonial and some post-colonial accounts of indigeneity (Sejersen, 2004). In contrast, the present analysis of Indigenous sense-making of nature taps into the collapse in the separation between man and nature and the insights that this offers into a post-growth economy (Brincker, 2017; Böhme, 1989).

First, the chapter presents the theoretical framework by drawing upon the scholarship of Karl Weick focusing especially on Weick's notion of sense-making. It then turns to the case and briefly introduces the reader to the setting and context, and furthermore to shadowing as a method. This brings us to the analysis, which focuses on two key elements of sense-making in the process of harvesting a polar bear: identity construction seen as a social and ongoing process, and the enactment of sensible environments. The chapter concludes by evaluating the potential of the study to contribute to our understanding of a post-growth economy.

The Theoretical Framework: Karl Weick's Theory of Sense-Making

The notion of sense-making is part of Weick's broader theoretical contribution to the study of organisations which hinges on three central concepts: enactment, sense-making, and heedfulness (often referred to as mindfulness) (Weick, 1995). The enactment theory states that when humans act, they bring structures and events into existence and set them in action. This forms the basis of a social-constructivist perspective on organisations. It offers a view that directs attention away from the organisation seen as an entity towards the continual processes through which organisations are brought to life. It is an approach where the analytical gaze of the observer shifts from the organisation to organising. The notion of sense-making is central to our understanding of organising, because it allows us to understand the process through which meaning is ascribed to experience. In doing so, it offers insights into what enables action.

Weick sees sense-making as a process that is grounded in identity construction, retrospective, enactive of sensible environments, social, ongoing, focused on and by extracted cues, and finally driven by plausibility rather than accuracy

(Weick, 1995, p. 17). These seven elements collectively shed light on the relationship between the individual and the environment. A key insight from Weick's work on sense-making is that shared meaning is not what is crucial in collective action. What is crucial is the experience of shared action. It follows that Weick does not work with a distinction between an individual mind and the collective mind. The collective mind precedes the individual mind. Hence, there can be no individual or single sense-maker (Weick & Roberts, 1993, p. 360). What a situation means is defined by who we become while dealing with the situation, and what or who we represent. This leads us to understand the individual as a person who is completely and fully embedded in the social, i.e. there is no individuality before social life.

In recent decades, the notion of sense-making has seen a revival, and it has been applied to studies that engage with organising under extreme conditions and circumstances—such as extreme weather and natural disasters (Whiteman & Cooper, 2011). Acknowledging that sense-making of ecological conditions is not a dominant concern within mainstream organisation literature, Whiteman and Cooper stress the need to embrace the power of materiality (Orlikowski, 2007; Suchman, 2005). They recognise that local inter-generational knowledge of ecosystems often enables people to make sense of and respond effectively to ecological change and surprise. However, ecological sense-making is not straightforward. The ability to make sense of nature, often at the spur of the moment, and respond sensibly requires that people are 'ecologically embedded'. According to Whiteman and Cooper, people who possess the quality of being ecologically embedded are those 'who are deeply rooted in "the land" in a physical and cultural way' (Whiteman & Cooper, 2011, p. 890). Those who do not have this characteristic are likely to suffer from its absence when they need to respond quickly and heedfully/mindfully to changes in the natural world. As a result, those who are not ecologically embedded and do not have ecological sense-making abilities become highly vulnerable when confronted with surprises that originate from the local ecology (Whiteman & Cooper, 2011, p. 891). With this in mind, the authors call for more research into acts and instances of sense-making of an ecologically material world.

Case and Context

Colony and County

Greenland, with its 2,166,000 square kilometres, is the largest island in the world. It is inhabited by approximately 55,000 people, who live along the coastal line with most settlements on the central west coast. Greenland is part of the North American continent. However, given that Greenland was a Danish colony for more than 200 years (1721–1953), a Danish county for more than 20 years (1953–1979), and is part of the Danish Realm, Greenland is also considered part of the Nordic countries, constituting the Arctic dimension. In recent years, the relationship between Greenland and Denmark and especially the colonial history has received a lot of attention among scholars and lay people alike. A consensus seems to be emerging that Danish colonial policies have been guided by ideas of protective paternalism and that these ideas have been facilitated by a unique bond between mission and trade. The policies effectively sealed off Greenland from the outside world and endorsed parts of traditional Inuit lifestyles that were seen as fit, while replacing elements that were regarded as unfit (Rud, 2014; Thisted, 2005, 2009, 2014).

Under pressure from the United Nations to allow the introduction of popular sovereignty, in 1953 Greenland's status was changed, and the island went from being a Danish colony to being a Danish county. In doing this, Greenland took a road very different from most other colonies that became independent. With the change in status from Danish colony to Danish county, it became a key priority for Denmark to assimilate Greenland into the Danish welfare state, which was created during the same period. The new policies had pronounced implications for the social structures in Greenland as the country went through a rapid modernisation process in areas such as health, housing, and education (Rud, 2014).

In recent years, the period during which Greenland was a Danish county has come under special scrutiny. This development has implied a far more critical understanding of the Danish presence and dominance in Greenland than what has been the case hitherto. In 2014, a reconciliation commission was launched by the government of Greenland. Interestingly, the Danish government declined to take part in the commission and the commission therefore became an exclusive Greenlandic activity. It became the main task of the commission to help span past, present, and future inside Greenland. The commission published its final report in 2017. The report identified five areas associated with the rapid process of modernisation that were seen as especially problematic in the sense that they had brought hardship and trauma to the people of Greenland (Andersen, 2019).

In 1979, as a response to a national and ethno-political movement that emerged in the 1960s and which was spurred on by the Danish dominance, Greenland acquired Home Rule. Its autonomy was further increased in 2009 with the Self-Government Act. The Self Government Act stands out by granting the government of Greenland the right to declare the country independent. It thus recognises the people of Greenland as a people with a right to self-determination. Furthermore, the Act allows the government of Greenland the right to assume legislative power over major policy areas. However, there are exceptions to this rule, most notably foreign and security policy (Brincker, 2022).

East Greenland is the most sparsely populated area of the territory with only 3000 inhabitants. Tasiilaq is, by Greenlandic standards, the only large town in this vast geographical area and has approximately 2000 people living there. East Greenlanders speak a distinct dialect that qualifies as a language unique to the region. In addition to differing from West Greenland in terms of language, East Greenland also has a different cultural and a shorter colonial history. East

Greenland was colonised in the late 1880s, while the rest of the country was colonised much earlier in 1721. This left East Greenland relatively more isolated than West Greenland. Also the county years 1953–1979 were experienced differently in East Greenland compared to the western region. Denmark's attempt to modernise Greenland and assimilate Greenland into the Danish welfare state focused primarily on West Greenland. As a result, Indigenous practices in hunting and fishing are relatively more prevalent and dominant in East Greenland. This extends to practices for hunting and harvesting polar bears.

Indigeneity and Isolation

Greenland belongs to the Inuit people, who are recognised as an Indigenous people. In contrast to many Indigenous peoples, Greenlandic Inuit constitutes the majority population of Greenland, and the majority of the population speaks the Inuit language, Greenlandic.

For many years, the Greenlandic Inuit have expressed hopes to be fully independent from Denmark one day, thus leaving the Danish Realm. A main obstacle to achieve this end derives from the fact that Greenland is economically dependent on Denmark. The Danish annual block grant constitutes about 50 per cent of the country's national budget. The hope to declare itself independent and to become the first Inuit nation state taps into the concept of Indigenous people and how to define it.

Attempts have been made to offer a definition of 'Indigenous people'. These efforts share the emphasis on the non-dominant position being vital to our understanding of this particular group of people (Mortensen & Barten, 2016). As Greenland gradually gains more autonomy and detaches itself from its former colonial ruler Denmark, a question emerges: Do the people of Greenland continue to be non-dominant and qualify as an Indigenous people? Furthermore, given the vast regional differences in Greenland that involve culture, language, and socio-economics, it is possible that some parts of the Greenlandic population are non-dominant (and thus Indigenous) whereas others are not. Domination and non-domination are fundamental to our understanding of indigeneity. If the former colonial ruler, i.e. Denmark, is taken out of the equation, and with relations of domination/non-domination playing out internally in Greenland, are the people of Greenland still an Indigenous people? Or may it be that when judged by the token of dominance/non-dominance, only parts of the population constitute an Indigenous people?

These questions are by no means just academic. In the early 2000s, the people living in Thule, in the northern part of the country, took legal action to be recognised as an Indigenous people distinct from the people of Greenland. The Danish Supreme Court ruled that the people of Greenland constitute one people, specifically one Indigenous people. However, the hope continues to linger, spurred on by the movement towards independence, and it is not limited to the Thule population. It is also present in East Greenland. The question therefore remains: At what point does the balance tip from being non-dominant to being dominant? Furthermore, when it tips, does it do so for the entire population of Greenland or does Greenland contain several peoples, some being Indigenous in the sense of being non-dominant, while others are not (Mortensen & Barten, 2016)? In the present context, it is vital to recall that in contrast to 'a people' or 'a minority', Indigenous people have certain rights relating to self-determination and to living and non-living natural resources (Mortensen & Barten, 2016). These include the rights of Indigenous people to hunt, for instance, seals and polar bears.

Methods: Shadowing in East Greenland

The motivation for carrying out research in one of the most remote and isolated parts of Greenland stems from a research project that the author conducted into Indigenous entrepreneurship and education in East Greenland (2016–2018).

The study took a point of departure in Barbara Czarniawska's method of shadowing. However, the researchers used it in a diluted form as we did not spend extensive periods of time (6–12 months) in the field. Instead, we had regular shorter stays. Shadowing was reported in field notes and narratives. Even though we used shadowing in a diluted form, Czarniawska's approach remained the central source of inspiration throughout the research.

The notion of mobility lies at the core of Barbara Czarniawska's work on shadowing as a method to study 'the work and life of mobile people' (Czarniawska, 2007, p. 17). Czarniawska calls for a mobile ethnography of people who move often and quickly. Consequently, she advocates not to follow the actor but to engage with action nets of actors that are under constant change (Czarniawska, 2004; Latour, 2007). Czarniawska makes a point of the fact that shadowing is more than a technique. It is a mindset that offers different insights. Shadowing requires that the researcher take a step back in order to observe. Ironically, in order to take this step back, the researcher has to step into the field and become part of it. Only then, according to Czarniawska, does observation in the true manner of shadowing become possible (Czarniawska, 2007, p. 9).

This raises the question of how the observed responds to the more or less constant presence of the observer. Based on the experience derived from the present study, the observed generally tend to ignore the observer after a very short period of time. While being ignored as an observer might be very attractive in the context of shadowing, this should not be considered an indication of 'going native', or an invitation to do so. Stepping into the field and going native are two very different things. The former allows the researcher to observe and capture that which escapes the eye of the observed. The latter signifies the point at which the researcher has spent so much time in the field that he/she begins to take things, events, and comments for granted (Czarniawska, 2007, pp. 27–28).

The above considerations are particularly relevant in the context of research on/with Indigenous peoples, especially given the context of colonisation. The author of this chapter is a Danish national and thus, in the eyes of the local community, associated with Denmark, the former colonial ruler. It raises the question of whether non-Indigenous researchers in general and nationals from the former colonial power in particular can be involved in accurate Indigenous research (Smith, 2012). For this study, the attempt was to avoid two pitfalls. One was the risk of producing research that ascribes a particularly privileged position to an underprivileged group, calling for a mobilisation for and protection of the group's interests, i.e. so-called standpoint theorists (ibid.). The other was not to engage in research that could potentially endanger the cultural particularity of the Indigenous group in question by intruding in the community. With this in mind, shadowing lends itself as a highly conducive method because it allows the researcher to balance stepping into the field and remaining distant from it at the same time.

How to Harvest a Polar Bear

The population of polar bears is not evenly distributed in Greenland. The vast majority can be found in the northern and eastern parts of Greenland. During the summer, polar bears follow their prey such as seals through the east coast of Greenland. Summer is, in other words, the season for hunting polar bears.

The Inuit people in the northern and eastern parts of Greenland have hunted polar bears for centuries. In addition to seal and fish, polar bears constitute part of the Inuit diet. To this particular group, hunting and harvesting polar bears is embedded in their Indigenous way of life and in their worldview.

There are specific traditions connected to hunting and harvesting polar bears. For instance, the person who spots the bear first has the right to the animal's skin. This is fixed. Consequently, if a child, an elderly person, or a woman spots the bear first, the skin is theirs, even though they do not participate in hunting the polar bear. Typically, however, it is a hunter that spots the bear first—as polar bears rarely come near villages or towns.

Sharing the bear's meat is guided by tradition, and hunters try to observe these traditions even today. Typically, the hunter who spots the bear first also shoots it. Apart from the skin, which is the reward he may claim for having spotted the bear first, he gets the head, meat from the chest bone, and meat from the central part of the bear, which is measured as the piece from the fifth rib from the top to the rib fifth from the bottom. If the bear is killed as part of a hunting expedition, the bear's limbs are divided equally among the remaining hunters. The intestines, which are considered a delicacy and an important source of nutrition, are divided among all the participating hunters (Møller, 2007).

Interestingly, if hunters who have not participated in the hunt happen to pass by, they also get their share of meat. They may claim the meat from the hip and the tail. Sometimes this piece of meat is also taken back to the village or town and offered to an old hunter who is no longer able to participate in hunting expeditions. This is considered a sign of respect (Møller, 2007).

The rule that the meat is shared both among participating hunters and hunters passing by is not limited to the polar bear. It applies to all types of large prey. For instance, in the case of narwhals, the tradition is that boats that happen to pass by the hunting scene may toss a stone or a cartridge on to the buoy that holds the whale afloat and thus claim a piece of meat from the belly of the whale. As we can see, the traditions that guide the sharing of prey are not limited to polar bears. They are also not limited to East Greenland. Similar traditions exist in Northern Greenland, albeit with local variations.

Sense-Making and How to Harvest a Polar Bear

The first insight to be derived from Indigenous lessons on how to hunt and harvest a polar bear pertains to the interaction between the individual and the collective mind. In this context, the understanding that there can be no individual sense-maker is vital. The collective mind precedes the individual mind. What a situation means is defined by who we become while dealing with the situation—such as the person who first spotted the bear or the former hunter who can no longer hunt. This process of identity construction is social and ongoing.

Against this backdrop, Indigenous practices connected to harvesting a polar bear involve an element of governance. Key to this governance is preventing conflicts arising among hunters over dividing and sharing scarce resources. For this reason, they clarify who gets the skin and the head of the bear, the two most precious parts of the animal, and who is entitled to what portion of meat. Governance also pertains to external relations, e.g. to passersby, because it clarifies that they have a right to claim a piece from the less attractive part of the animal, the hip.

When engaging with Indigenous practices as movement and patterning of social form rather than as system or structure, these social practices are linked to questions of organising, and processes of sense-making that give meaning to organising. Hence, they allow us to understand the significance of these elements of governance in the context of a post-growth economy. The aim of governance is to prevent conflict arising over scarce resources.

The social and ongoing process of identity construction is also clearly expressed in the granting of rewards. Indigenous lessons on how to harvest a polar bear include insights into rewards that relate to the risk associated with spotting and shooting a polar bear. For example, the hunter who does so gets the skin and head—parts of the animal that are considered most attractive—as well as a significant portion of the meat. However, rewards extend to the entire group and to passersby as well. Hence, while they are granted to particular individuals, they do not cater to an underlying understanding of individuality. Instead, these lessons are embedded in identity construction that acknowledges and seeks to promote actions and rewards risk-taking in the context of human survival, not as individuals but as species living along with other species in a space that is social and where resources are scarce.

As mentioned, the system of rewards extends to former hunters who no longer have the capacity to hunt. At first glance, this may appear to be a social welfare scheme. However, the scheme is limited to retired hunters; it does not cover all elderly people in the group. So instead of envisioning this as a welfare scheme, it is more apt to perceive this particular reward system as an old-age insurance scheme targeted towards former hunters.

The distinction between social welfare and an insurance scheme is very important in the present context, as it taps into the rationale that informs this system. In the case of social welfare, the system is informed by rights and duties that are assigned to certain individuals. In the case of an insurance scheme, the rationale that underpins the system can be described as an act that motivates a particular type of behaviour over time. The hunters who bring meat back to the old hunter do so anticipating that young hunters will, in the future, do the same for them when they are too old to go hunting. The act involves an expectation of reciprocity (Elster, 1992).

The distinction between a social welfare regime and an insurance scheme is significant for the discussion on individuality and its place—or lack thereof—in Indigenous lessons on how to harvest a polar bear. When exploring the act of bringing part of the prey back to former hunters, it becomes clear that this act relies on neither altruism nor self-interest. Instead, it is an act that spans and links the past, the present, and the future and is grounded in the social and ongoing process of identity construction.

Finally, an insight to be derived from Indigenous practices connected to the hunting and harvesting of polar bears relates to the environment, specifically Weick's notion of sense-making in relation to the material world: the enactment of sensible environments. The enactment of sensible environments implies that, following Weick, people produce the environment that they face. There is no fixed environment, according to Weick. People are part of their own environments, which they produce by acting in them (Weick, 1995, p. 31). In saying this, Weick is not implying that there is no material world outside our cognition. Only that when we focus on it, we enact it and construct it socially. In doing so, we construct sensible environments. This extends to the enactment of the natural environment which Whiteman and Cooper seek to capture with the notion of ecological sense-making (Whiteman & Cooper, 2011). In the case of Indigenous lessons on how to harvest a polar bear, sense-making of the natural environment or ecological sense-making means that the polar bear is enacted as prey, an animal to be hunted and harvested for food. It also means that other enactments are possible. One such example is from West Greenland, where the polar bear is enacted as a sacred animal and a national symbol. This is clearly expressed in the fact that a roaring polar bear is a national symbol of Greenland. Another is in the context of the wider world where images of starving polar bears are typically enacted as symbols of environmental degradation stemming from the economic growth of the western world.

The above-mentioned three enactments of the polar bear as prey, sacred animal and national symbol, and finally a symbol of environmental degradation are part of cognition that involves sustainability. Among the Indigenous people living in East Greenland, sustainability is the sustainability of humans placed in a nature that is socially constructed in the ways that the livelihood of living resources such as the polar bear is fundamental for the sustainability of humans. The enactment of sensible environments involves the sustainability of humans as a species, as opposed to the sustainability and self-preservation of sovereign individuals.

Discussion and Conclusion

This chapter set out to explore how Indigenous lessons on how to harvest a polar bear offer insight into sense-making in a post-growth economy. Taking its inspiration in Czarniawska's method of shadowing, her call for a mobile ethnography, and the concomitant point to follow the action rather than the actor, the chapter integrated research conducted into entrepreneurship and education at a state school in East Greenland with an analysis of the social practices involved in the hunting and harvesting of polar bears in that region. In exploring and unfolding these practices, the chapter has drawn on Karl Weick's theory of sense-making, in particular the process of identity construction and the enactment of sensible environments.

The chapter has identified elements of governance and rewards that underpin how the meat and the skin of the bear is divided both internally within the group of hunters and in relation to passersby. It has discussed the rationale that informs rewards, illustrating how they seek to motivate particular actions such as risk-taking. Rewards are not limited to the present moment but extend to former hunters and to the future; they involve an expectation of reciprocation and are firmly embedded in the collective mind.

The question that remains to be addressed is how Indigenous lessons on how to harvest a polar bear offer insight into sense-making in a post-growth economy. This chapter argues that Indigenous practices can facilitate such insights by enabling us to envision commons that are profoundly social in their conceptualisation. Relationality is key to this understanding. When engaging with Indigenous practices as movement and patterning of social form rather than as system or structure, these social practices are linked to questions of organising, and processes of sense-making that give meaning to organising. At the heart of this notion is the recognition that the collective mind precedes the individual mind.

Weick's works are dedicated to analysing processes of sense-making, including how people enact their own environment—both social and natural. To the extent that it makes sense to speak of sustainability in relation to this collective, it pertains to humans as a species, and not to sovereign individuals. The question is one of not becoming extinct as a species as opposed to self-preservation of the individual.

References

Andersen, A. N. (2019). The Greenland Reconciliation Commission: Moving away from a legal framework. *The Yearbook of Polar Law, XI*, 214–244.

Baker, G. (2019). Critique, use and world in Giorgio Agamben's genealogy of government. *Global Society, 33*(1), 12–25.

Bauhardt, C. (2014). Solutions to the crisis? The Green New Deal, Degrowth, and the Solidarity Economy: Alternatives to the capitalist growth economy from an ecofeminist economics perspective. *Ecological Economics, 102*, 60–68.

Böhme, G. (1989). *Für eine ökologische Naturaesthetik*. Suhrkamp Verlag.

Bonneuil, C., & Fressoz, J. (2017). *The shock of the anthropocene*. Verso.

Brincker, B. (2017). Images of the arctic: Visualizing Greenland as an indigenous people and a modern nation. *Visual Studies., 32*(3), 251–261.

Brincker, B. (2022). Greenland: Autonomy in the Arctic Region. In B. Hong & A. Ichijo (Eds.), *The Routledge handbook of territorial autonomies*. Routledge.

Chandler D. (2018). The death of hope: Affirmation in the anthropocene. *Globalizations, 15*, 695–706.

Chandler, D. (2019). The transvaluation of critique in the Anthropocene. *Global Society, 33*(1), 26–44.

Colebrooke, C. (2014). *Death of the PostHuman: Essays on extinction* (Vol. 1). Open Humanities Press.

Czarniawska, B. (2004). On time, space and action nets. *Organization, 11*(6), 777–795.

Czarniawska, B. (2005). Karl Weick: Concepts, style and reflection. *Sociological Review, 53*(1), 267–278.

Czarniawska, B. (2007). *Shadowing and other techniques for doing fieldwork in modern societies*. Liber.

Elster, J. (1992). *Local justice: How institutions allocate scarce goods and necessary burdens*. Russell Sage Foundation.

Jack, G., Westwood, R., Srinivas, N., & Sardar, Z. (2011). Deepening, Broadening and Reasserting a Post-colonial interrogative space in Organization Studies. *Organization, 18*(3), 275–302.

Latour, B. (2007). *Reassembling the social: An introduction to actor-network-theory*. Oxford University Press.

Møller, L. P. A. (2007). *Fangerne fra Ammarsalik*. Det Grønlandske Hus.

Mortensen, B. O. G., & Barten, U. (2016). The Greenland Self-Government Act: The pitfall for the Inuit in Greenland to remain an indigenous people. *Polar Law Yearbook, 8*, 103–129.

Orlikowski, W. J. (2007). Sociomaterial practices: Exploring technology at work. *Organisation Studies, 28*, 1435–1448.

Rud, S. (2014). Governance and tradition in nineteenth century Greenland. *Interventions: International Journal of Postcolonial Studies, 16*(4), 551–571.

Sandberg, M., Klockars, K., & Wilen, K. (2019). Green growth or degrowth? Assessing the normative justifications for environmental sustainability and economic growth through critical social theory. *Journal of Cleaner Production, 206*(1), 133–141.

Sejersen, F. (2004). Horizons of sustainability in Greenland: Inuit landscapes of memory and vision. *Arctic Anthropology, 41*(1), 71–78.

Smith, L. T. (2012). *Decolonizing methodologies: Research and indigenous peoples* (2nd ed.). Zed Books.

Suchman, L. (2005). Affiliative objects. *Organization, 12*(3), 379–399.

Thisted, K. (2005). Postkolonialisme i nordisk perspektiv: relationen mellem Danmark-Grønland. In H. Bech (Ed.), *Kultur på kryds og tværs*. Klim.

Thisted, K. (2009). Where once Dannebrog waved for more than 200 years: Banal nationalism, narrative templates and post-colonial Melancholia. *Review of Development and Change, 16*(1–2), 147–172.

Thisted, K. (2014). Imperial ghosts in the North Atlantic: Old and new narratives about the colonial relationship between Greenland and Denmark. In D. Góttsche & A. Dunker (Eds.), *(Post) Colonialism across Europe: Transcultural history and national memory*. Aisthesis Verlag.

Weick, K. E. (1995). *Sense-making in organizations*. Sage.

Weick, K. E., & Roberts, K. E. (1993). Collective mind in organization: Heedful inter-relating on flight deks. *Administrative Science Quarterly, 38*(3), 357–381.

Weick, K. E., Sutcliffe, K. M., & Obstfield, D. (2005). Organizing and the process of Sense-making. *Organization Science, 16*, 409–421.

Whiteman, G., & Cooper, W. H. (2011). Ecological sense-making. *Academy of Management Journal, 54*(5), 889–911.

Xu, Q. (2008). A question concerning subject in the spirit of Chinese capitalism. *Critical Perspectives on International Business, 4*(2–3), 242–276.

CHAPTER 9

Wind Power, the EU (Un)Green Deal, SDG7, and Environmentally Destructive Settler Colonialism in Indigenous Sámi Territories: Hällberget

Elle Eriksson and May-Britt Öhman

Introduction

Energy production often affects traditional territories, where Indigenous peoples have lived for millennia (Whyte, 2023). This is also the case regarding wind power in Northern Europe. Wind power plants are increasingly being built on Indigenous Sámi reindeer herding territories, despite the negative impacts on reindeer herding shown in several studies (Cambou et al., 2021).

Despite its dependency on fossil fuels for construction, maintenance, and decommissioning, and despite several other negative environmental impacts, wind power is repeatedly presented as a renewable, fossil-free, and sustainable energy source. It has been one of the fastest growing sources of so-called green energy worldwide in recent years. In 2022, 37.5% of EU-generated electricity came from wind power (Eurostat, 2024), and it is an important pillar within the European Green Deal, a programme for growth adopted by the European commission in 2019. The Green Deal is intended as a way to put the EU on the path towards a "green transition", with the goal of reaching climate neutrality by 2050. This is the EU contribution to the Paris Agreement, the internationally binding treaty to limit global warming and mitigate climate change

E. Eriksson (✉)
Swedish University of Agricultural Sciences, Umeå, Sweden
e-mail: elle.eriksson@slu.se

M.-B. Öhman
Uppsala University, Uppsala, Sweden
e-mail: may-britt.ohman@cemfor.uu.se

R. Toivanen et al. (eds.), *Decolonizing the Sustainable Development Goals*,
https://doi.org/10.1007/978-3-032-14263-4_9

adopted at the UN Climate Change Conference (COP21) in Paris in 2015 (Council of the EU, 2024) (Fig. 9.1).

Sustainable Development Goal 7 (SDG7) of the UN's 2030 Agenda for Sustainable Development, adopted by the UN General Assembly in 2015, has a similar direction, the aim being to ensure an affordable, reliable, and just energy transition for all people on Earth (UNGA, 2015).

The EU Green Deal makes no mention of minorities or Indigenous peoples. However, the EU's founding treaty stipulates that member states shall respect cultural diversity, combat discrimination, promote social justice, and uphold the rule of law and human rights, including the rights of persons belonging to minorities (European Union, 2024). Furthermore, when Sweden and Finland entered the EU, Swedish negotiators made sure to have Sámi reindeer herding rights enshrined in the accession documents, granting "exclusive rights to reindeer husbandry within traditional Sámi areas" (Laframboise, 2023).

Meanwhile, UN Agenda 2030 commits to protecting human rights, empowering vulnerable groups—including Indigenous peoples—and ensuring the lasting protection of the planet and its natural resources (UNGA, 2015).

Indigenous peoples worldwide, despite decades of intense and continuous efforts within national and international organizations, still remain largely marginalized when it comes to decision making over lands and natural resources (Whyte, 2023). Sámi demands for recognition and respect have persisted for over a century (Laula, 1904; Lantto & Mörkenstam, 2008). Yet, Sámi are still today left out from the spaces, such as courts and political arenas, where

Fig. 9.1 A smaller herd of reindeer inside a traditional timber corral, located approximately three kilometres from the planned wind power area, Hällberget. Herders keep the reindeer in the corral for as short a time as possible, ranging from a few hours to a few days, since the risk of catching diseases increases the longer they stay in the corral. (Photo: Elle Eriksson)

decisions that have a major impact on their livelihood are taken. This happens despite the fact that the Sámi rights to reindeer husbandry are enshrined in the Swedish constitution, and that legislation stipulates that land owners and other stakeholders of lands within the reindeer husbandry area must safeguard and respect their grazing rights (SFS 1974:152, 2 kap. 17 §; SFS 1971:437, 94 §; Cambou et al., 2021).

The main aim of this chapter is to discuss Sustainable Development Goal 7 (SDG7) and the EU Green Deal from a decolonizing perspective, challenging the false promise of a sustainable and green transition through wind power establishments, through acknowledging Indigenous Sámi knowledges, rights, cultures, and values, in particular in regard to reindeer husbandry.

This is done by presenting a case study, based on interviews and personal communications, film and photo documentation, storytelling, and notes from own observations on how planned wind power plants affect and will affect the Váhtjer (Gällivare) Forest Sámi community in northern Sweden, and in particular the Várre-group (*sijdda*). The case study is contextualized with special regard to today's reindeer herding practice. Our methodologies and theories are based on the disciplines of environmental history, history of technology, Árbbediehto—Sámi traditional knowledge (Nordin Jonsson, 2010)—and constructivist grounded theory (Charmaz, 2017).

Elle Eriksson is a member of the Váhtjer Forest Sámi community and the Várre-group, and a PhD student. She thus has both lived experience and theoretical knowledge of the current case study. May-Britt Öhman is Lule and Forest Sámi of the Lule River valley, with Tornedalian heritage, and an associate professor of environmental history, and has collaborated with, followed, and documented the struggles of the Várre reindeer herders for more than a decade (Fig. 9.2)

Settler Colonialism, Indigenous Peoples, the Sámi, and the Forest Sámi

According to the United Nations (n.d.), Indigenous peoples amount to around 480 million, or around 6% of the world's population, in around 90 states. Indigenous peoples steward or hold tenure rights over more than one-quarter of Earth's terrestrial surface, representing at least 37% of lands with a low level of ecological disturbances (Fernández-Llamazares et al., 2024).

Indigenous peoples are recognized as having strong affiliations with their traditional territories, pre-settler livelihood, and the upholding of distinct social, cultural, and political organizations as well as languages, beliefs, and knowledge systems. A common denominator for Indigenous peoples is that their land use tenure and rights have been, and still are being, violated, under an ongoing process of settler colonialism. In several cases, settler colonial power holders are promoting the elimination of Indigenous peoples through assimilation or in some cases homicide, dispossession, and displacements from

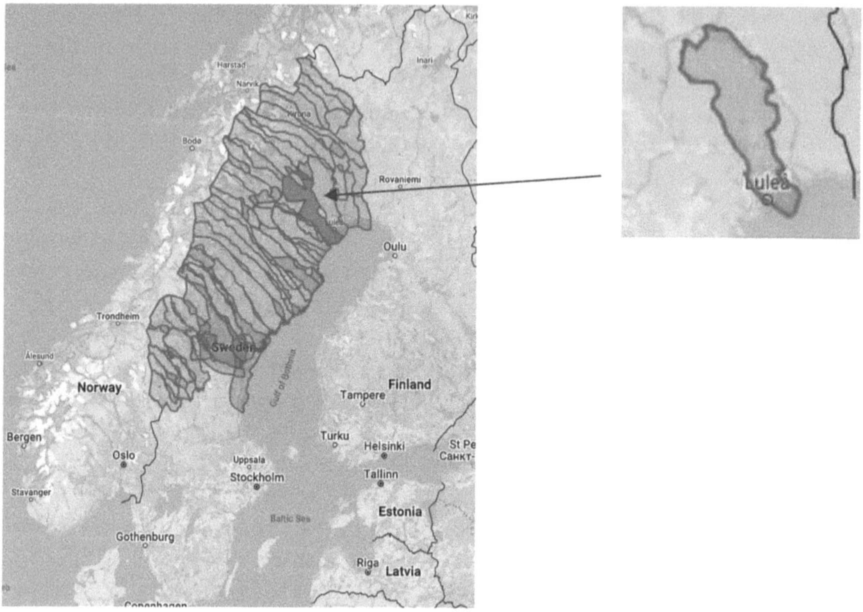

Fig. 9.2 Reindeer herding communities—*samebyar*—and the current reindeer herding area on the Swedish side of Sábme. Google GeoBasis-DE/BKG (© 2009) via Sametinget. 9.2.2. Váhtjer (Gällivare) Forest Sámi territories of today, Google GeoBasis-DE/BKG (© 2009) via Sametinget

their territories, and thereby the destruction of Indigenous peoples' relationships to lands and waters, and access to their traditional food sources (O'Brien, 2017; Wilson & Laing, 2019; Whyte, 2023).

Relations with reindeer and reindeer herding have since millennia been a precondition for human livelihood in the Arctic region, and are still upheld by more than 20 different Indigenous peoples (Mathiesen, 2023). Since the last ice age, the Sámi have lived across the entire Fennoscandian Peninsula. The Sámi peoples' territories are in the Sámi language called Sábme, stretching across four settler colonial nation states: Sweden, Norway, Finland, and the northwestern part of Russia. Similar to other Indigenous territories, Sábme is a complex system of interacting sociocultural and ecological relations. Sámi have traditionally sustained themselves by fishing, hunting, gathering plants, and/or herding reindeer, as well as cultivating, tar making, and trade. The Sámi were also known to be ship builders and smiths, and have also upheld high positions in society as vicars, sextons, teachers, members of court, and bailiffs (Lanyon-Garrido et al., 2024; Larsson, 2021; Larsson & Päiviö Sjaunja, 2022, p. 71FF; Öhman, 2021; Bennerhag & Söderholm, 2024).

During the nineteenth century, parallel with industrialization and quest for natural resources in the northern regions, Sámi territories were increasingly colonized by the Swedish state, while Sámi were depicted and treated as different and as an inferior race from the Nordic. However, Sámi have via mobilization, on a national level and international level, gained certain influence, status, and recognition. In 1977 the Sámi were recognized as a People within a Swedish Government inquiry. With the 1994 Council of Europe Framework Convention for the Protection of National Minorities, the Sámi were acknowledged as one of five National Minorities in Sweden and the Sámi languages were recognized as National Minority languages (Öhman, 2024; Lantto & Mörkenstam, 2008). In 2011, the Sámi were recognized as a People within the Swedish constitution (SFS 1974:152) and are also recognized as the only Indigenous peoples on the European continent (Laframboise, 2023).

The reindeer herding Sámi communities (*samebyar*) are also rights holders within their territories, recognized within Swedish legislation since the first reindeer herding act in 1886 (SFS 1971:437). This right is based on their earlier ownership of lands, the Sámi tax lands, further described below. While the Swedish state collectivized the reindeer husbandry organization into larger Sámi communities, the family-based groups or *sijdda* (*siida* in North Sámi) system was and remains the core organizational form (Labba, 2016).

Today, around 50% of Sweden's total land area is reindeer husbandry territory, although very little of this is usable for reindeer grazing. Since 1971 the territory in Sweden has been divided into 51 Sámi communities. Two main forms of reindeer herding are practised, mountain and forest reindeer herding. Simplified, mountain reindeer herders practise reindeer herding in the mountains during the summer and migrate to forest landscapes in the winter. Forest reindeer herders remain with the reindeer in the forest landscapes through all seasons. There are 10 forest Sámi communities in Sweden, nine of which are located in Norrbotten County (Sametinget, 2024).

In the late nineteenth and early twentieth centuries the Forest Sámi reindeer herders were not considered by settler colonial Swedish authorities as "authentic" Sámi, and were therefore subject to discrimination and loss of grazing lands. However, Forest Sámi spokespersons, amongst others Karin Stenberg (1884–1969), have been speaking up for Forest Sámi rights and culture (SBKL, 2018; Larsson, 2021; Öhman, 2024) (Fig. 9.3).

Sámi Experiences of Land-Use Pressures

From the early 1900s and throughout the entire twentieth century, the state exploited most of the major rivers within Sámi reindeer herding territories. This led to massive loss of grazing and calving lands as well as reindeer migration routes. The consequences still remain (Össbo, 2021). Since the late nineteenth century, power lines, mining, industrial forestry, roads, railways, expanding cities, increasing tourism, and military bases have all led to further environmental destruction and with negative impacts on Sámi reindeer herding, fishery, culture, livelihood, and wellbeing. By 2022, 85% of northern Fennoscandinavia was affected by at least one land-use pressure and 60% were impacted by multiple pressures. Consequently, the majority of reindeer grazing

Fig. 9.3 Reindeer looking for good winter food, lichen, at Dockasberg, less than one kilometre from the nearest planned Vasa Vind wind power site. March 1, 2025. (Photo: Henrik Andersson)

lands are thus subject to cumulative encroachment pressure, which results in fragmentation and deterioration of grazing resources, a major concern within the Sámi society (Stoessel et al., 2022; Axelsson-Linkowski et al., 2020).

Since the early twenty-first century, wind power has become the new threat to Sámi livelihood and reindeer husbandry. Areas which are more heavily populated and thus have a higher demand for energy, in particular near the cities in the south of Sweden, do not experience the same degree of wind power establishment as Sámi reindeer herding territories (Energimyndigheten. 2024). A reason for this may be that in the more populated areas in the south of Sweden, local inhabitants protest wind power establishments in their vicinity and the municipalities in Sweden have since 2009 a right to veto wind power establishments. In the north, municipalities which despite richness in natural resources have declining populations tend to be more positive as local politicians are hoping for economic benefits for the local community (Lindvall, 2023).

There are several studies showing the negative impact on reindeer herding. According to Skarin et al. (2021), both reindeer and the reindeer herding practice are negatively affected by wind power development. Skarin et al. (2021) found that during the calving, summer, and autumn periods of the bare ground season, the reindeer within the Forest Sámi community of Malå were negatively affected by the establishment of wind power plants and related infrastructure. The impact was most severe in the important mire areas where the reindeer's usage of the grounds decreased by on average 34% in the calving period and 21% in the autumn period. These effects were observed up to three kilometres away from the wind power plants.

Other environmental impacts and disturbances caused by wind power include, amongst other things, the quarries that are needed to construct the access roads for wind power plants. These are placed near the sites, leading to loss of reindeer lands. The many access roads with heavy traffic lead to disturbances (Öhman et al., 2021). Furthermore, wind power plants require concrete, steel, iron, fibreglass, polymers, aluminium, copper, zinc, and rare earth metals, leading to an increased demand for mining. The majority of mines in Sweden are located in reindeer herding territories and the green transition is used as an argument for promoting even more mines (Alves Dias et al., 2020; IEA, 2021).

Recent studies also indicate that the erosion of wind turbine blades caused by rain, hail, and wind results in the release of microplastics and chemical contaminants, per and polyfluoroalkyl substances (PFAS). PFAS and microplastics have been observed to contaminate groundwater, surface water, and soil and may cause long-term harm to human health and to the environment (European Chemicals Agency, n.d.; Mishnaevsky Jr. et al., 2024; Keating, 2023). So far, there are no studies performed within reindeer grazing lands that have measured PFAS and microplastic contamination of soils and waters.

Hällberget in Váhtjer Sámi Community

The Forest Sámi Váhtjer community territory is about 8300 km^2 and the maximum number of reindeer allowed by the Swedish state is 7000.

Today there are around 35 active reindeer herders in the community that, usually with the help of their family, carry out reindeer herding. Reindeer herders have always adapted the herding according to prevailing conditions but during the two latest decades the changes have been drastic, with increasing encroachments from industrial forestry, mining, increasing traffic, ski-races, car rallies, power line constructions, and wind power (Eriksson, 2021; Öhman et al., 2021; Mumford, 2021; Reindeer herder, Personal communication, Feb. 2025).

Since about a decade, two major wind power establishments are in the planning in the Váhtjer community area: Storlandet by the Swedish state power company Vattenfall, and Hällberget by the company Vasa Vind, owned by the Dutch pension fund APG (Mumford, 2021; Vasa Vind, n.d.).

Hällberget, in Lule Sámi called Hällovárre, is one of the main calving areas in the Váhtjer community. Part of an earlier Sámi tax land, it is a traditional territory for the Várre-group. The Várre-group hold the rights for reindeer herding here, and has utilized the area for centuries.

In 2014, Vasa Vind applied to the Environmental Assessment Delegation (EAD)—miljöprövningsdelegationen—within the County Administrative Board of Norrbotten County to be granted permission to establish a group station for wind power with a maximum of 150 wind turbines. The station would have a "box-model" with associated facilities, roads, and external infrastructure (MMD, 2021). During the process within the EAD, the application was revised to include a total of 123 wind turbines, each positioned within a circular patch with a radius of 200 metres.

The wind power is planned to be established in connection to and in the surroundings of Hällberget, which is classified as both a nature reserve and a Natura 2000 area.[1] Since the state grabbed the lands area from the Sámi in the nineteenth century, the area is mainly owned by Sveaskog, the Swedish state-owned forest company. There are also smaller parts therein which are owned by private forest owners. The applied area covers 90 km^2 and within the area there are two designated zones of so-called national interest (*riksintresse*) for wind power,[2] both in the northern part. However, there are also five designated areas of national interest for reindeer husbandry located between 1.5 and 10 km from the planned area. These are all situated on traditional territory for the Váhtjer community. The closest national interest for reindeer husbandry is the calving grounds for the Várre-group, which is only about 1.5 km northwest of the applied area. It was designated a national interest for reindeer herding because of its status as a calving area, but also since the available grazing is good both in the summer and winter (MMD, 2021). Another area of national interest, due to its cultural environment and nature conservation, is also located northwest of the applied area. In addition, there are source flows to the Natura 2000 area of Torne and Kalix river systems within the applied area, as well as three additional nature reserves nearby.

The Överkalix municipality politicians had the opportunity to safeguard the interest of the Sámi reindeer husbandry. Yet, despite the objections from the Sámi community, in 2015, an absolute majority, 24 out of 29 members of the municipal council, voted yes to the Vasa Vind proposal, claiming it would lead to economic development (Salman, 2015). Furthermore, the largest land owner, state-owned Sveaskog, should have consulted the Váhtjer community before granting land for wind power, but neglected their responsibility to do so (Naturskyddsföreningen, 2024).

In 2019, the EAD, despite the objections of the Sámi community, granted Vasa Vind permission to construct and operate wind turbines on a maximum of 57 patches. The permission included some limitations and protective measurements, for example regarding reindeer husbandry and birdlife. Vasa Vind, Váhtjer Sámi community, and the Swedish Society for Nature Conservation in Norrbotten County all appealed the EAD decision to the Land and Environment Court (Mark- och miljödomstolen (MMD)). Sweden's Ornithological Association–BirdLife Sweden and the Ornithological Society of Norrbotten submitted a joint appeal of the decision as well.

[1] Natura 2000 is a network of protected areas within the EU. These are natural areas with species or habitats that are considered particularly worthy of protection from a European perspective.

[2] The so-called national interests are regulated in the Swedish Environmental code (SFS nr: 1998:808), and are used to safeguard interests that are of major importance in land use planning, such as mining, national defence, reindeer herding, fishery, cultural heritage, biodiversity, and more recently also wind power.

The MMD (2021) court proceedings were held in April of 2021 and the verdict was declared two months later. Also here, 57 out of the 123 applied wind turbines were approved, while the designated operational area was reduced by approximately half.

Várre, Traditional Sámi Land Use, and Reindeer Herding at Hällberget

The Várre settlement was established in 1836 (Andersson, 2023, p. 7). Today, it is a village with a single permanent resident. However, the history of the Várre settlement and the Sámi affiliated with it begins long before the nineteenth century.

Numerous state interventions have targeted the Sámi over the last 500 years. One of the major interventions was the dispossession of Sámi tax lands, lands that Sámi families owned and controlled. The tax lands were generally relatively large, and they included various sizes of hunting grounds, grazing lands for reindeer, and fishing waters (Larsson & Päiviö Sjaunja, 2022). By the end of the eighteenth century, while Swedish peasants through a royal decree gained stronger rights to their lands, the Sámi were not provided the same protection. This was followed by a delineation process—*Avvittringen*—starting from the end of the eighteenth century and finalized by 1928. During this process the state took Sámi tax lands and gave these to farmers (of which some were Sámi who already had or had to give up reindeer herding; others were Swedish and Finnish). Some were kept as state property, today to a large extent managed by Sveaskog. Swedish state representatives depicted reindeer herding as far less valuable than farming, and deemed it a vanishing lifestyle (Andersson, 2023, p. 10; Öhman, 2024; Lantto & Mörkenstam, 2008; Larsson, 2021; Norstedt, 2018, p. 73f) (Fig. 9.4).

The northern parts of Sweden have since the fourteenth century been referred to as the Sámi lands, *Lappmarken*. The area was divided into several different *lappmarker*, Sámi districts (Larsson & Päiviö Sjaunja, 2022, p. 42).[3] Várre is located in Lule lappmark of the Lule Sámi language area, with the major watershed being the Lule River. Within Lule lappmark there was a vast forest that the Sámi community called Sjokksjokk, with several recorded Sámi tax lands (Andersson, 2023, p. 26). According to genealogy done by one of the Várre Sámi, the Várre family owned two tax lands: Maisavuoma and Tjorvek (Hornberg), including Wuottasråfve. However, due to the delineation process, their ancestors were forced to establish a new settlement at the end of the eighteenth century, which they named Häjmmaskájdde. The family lived there until 1836, when they moved approximately three kilometres away to establish another settlement, named Várre (Andersson, 2023, p. 7ff).

The reindeer that feed on lichens in winter and on mosses, herbs, ferns, grasses, and mushrooms during the other seasons have unique features adapted

[3] "Lapp" is an older outdated Swedish word for Sámi.

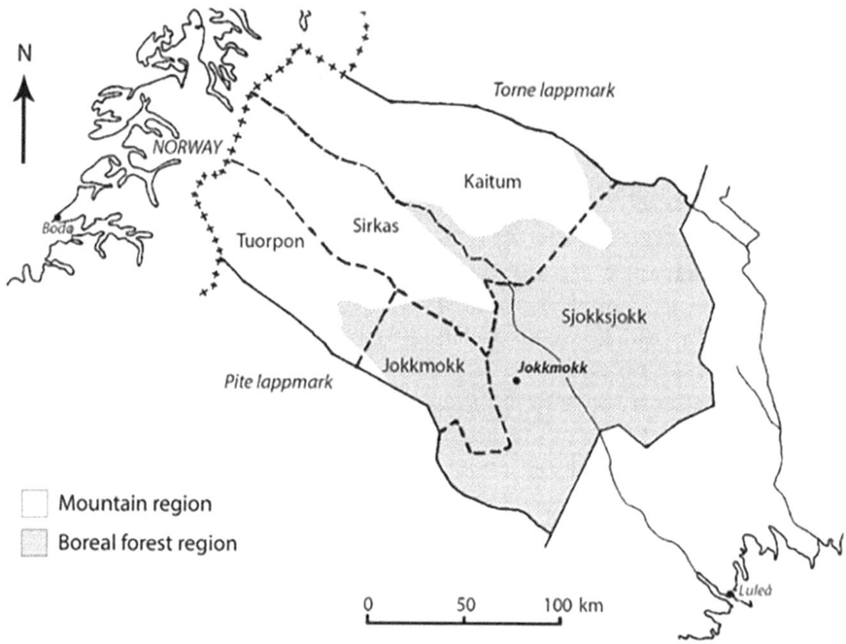

Fig. 9.4 Map of the Swedish state defined Lule lappmark circa 1760 with approximate borders between Sámi villages, Sjokksjokk and the others, and the *Lapland border* (Lappmarksgränsen) to the coastal areas (Larsson and Päiviö Sjaunja, 2022). The Lapland border was the result of a decades-long process and was established by the Swedish state in the mid-eighteenth century to regulate land use between the reindeer-herding Sámi and settlers west of the border and the farming communities in the coastal parishes. However, reindeer herding has long taken place east of the border, in the coastal areas and on the islands of the archipelago (Bergman, 2021). The border is, however, still used within legislation regulating reindeer herding (SFS 1971:437)

to the harsh Arctic climate. Reindeer have been critical for human survival and livelihood in the region. From hunting reindeer via the (semi-) domestication of reindeer, the Sámi have accumulated knowledge on the reindeer's ecology, and learned how to live with and from the reindeer (Inga, 2008).

The earliest written indication of domesticated reindeer is from 890 AD (Bjørklund, 2013). Larsson and Päiviö Sjaunja (2022) describe that the use of domesticated reindeer in interior northern Fennoscandia from its introduction to the present day can be divided into three major phases of successive tenure systems. In the first phase, most or all households had small numbers of domesticated reindeer that they used to carry loads and haul sledges, or as decoys when hunting wild reindeer. The reindeer were also used for milking and the oldest historical milking ground known today dates to at least the 1350s (Egelkraut et al., 2018). During this phase, hunting and fishing constituted the foundation of the household economy, although the domesticated reindeer provided many essential products and services to the household.

The second phase was labour intensive with comparatively large herds that had to be rounded up each day to be milked. Instead of hunting and fishing, the products from domesticated reindeer became the focus of the household economy. To ensure year-round grazing, most households moved between summer and winter grazing grounds in the landscape. During this phase, the reindeer husbandry can be characterized as pastoralism because of the moveable lifestyles of the households and since the reindeer fed on large, usually unfenced, grazing lands. In the third phase, from the nineteenth century, reindeer husbandry became more extensive, almost entirely focusing on meat production for sale. All these changes happened due to state interventions through taxation, land grabbing, and legislation. Reindeer pastoralists turned into reindeer herders (Larsson & Päiviö Sjaunja, 2022). Besides reindeer herding, also fishing, hunting, and collection of berries and mushrooms are still of major importance for many households (Lanyon et al., 2024).

Traces of Sámi presence and relationships with the lands are found both in historical records and in the landscape. Records show that the Várre Sámi have been practising intensive reindeer herding from at least the seventeenth century, although it might be earlier (Andersson, 2023). On the Várre-group's territory there is a sacrificial site named Unna Sájvva, which was excavated in 1915. The metal objects and coins have been dated to the tenth and eleventh centuries AD. However, the animal bone finds are older, some dating from the sixth century AD. The site is still of major importance (Andersson, 2024; Salmi et al., 2015). Blazing, removal of the outer bark on trees, was common and was done for several different purposes: marking borders, indicating hunting paths, marking trails, for religious purposes, and also to extract bark for food (Eidlitz Kuoljok, 1969). Such blazes are found in the Várre area.

Hence, there is a long continuity of Forest Sámi presence in the studied area, their traditional knowledge being passed on from generation to generation. Today the oldest of the Várre descendants is 95 years old. Currently, there are three generations of Várre Sámi with close connection to reindeer herding: one generation of Elders and two generations of active reindeer herders (Fig. 9.5).

Wind Power at Hällberget: A Threat to the Sámi Reindeer Husbandry Futures

The current industrial exploitations of the traditional Várre territories started in the 1950s. From the beginning of the 2010s, wind power was added to the long list of intrusions. As explained above, the court process between the Várre Sámi and Vasa Vind was ended in 2021 when the MMD decided to grant the permission for wind power. By 2025, the constructions had not yet started. In March 2024, Vasa Vind stated that they wished to apply for a change in the

Fig. 9.5 A dead old tree with a blaze—in this case a border marking—inside a reindeer corral located approximately 6 km from the planned wind power area. (Photo: Elle Eriksson)

permit, with one detail being to increase the height of the wind turbines from 200 to 250 m (Vasa Vind, 2024). However, by early 2025, the application had still not been made.

The threat from wind power, which by now has been ongoing for more than a decade, has had a large impact on the whole community, and has led to worry and anxiety for what the future holds. The Várre-group has had to seek legal advice to promote their perspectives. The young generations, who have grown up with this threat from childhood, doubt whether they can continue with reindeer herding, and this has had a large negative impact on mental wellbeing (Storlöpare et al., 2024; Andersson et al., 2019).

With the constructions, which may last for several years, increased traffic will be a major problem. It is likely that the number of reindeer run over by trucks will increase. They also increase the risk of accidents when the reindeer herders perform their everyday work among the reindeer, as they need to share the road with the construction traffic. Since the fall of 2024 the Várre-group have been experiencing the ongoing work with the Aurora power line —from the Messaure hydropower station on the Lule River to Finland, run by the Swedish state authority, Svenska kraftnät, and promoted as part of the green transition

Fig. 9.6 View from Hällberget, with its primeval forest. While this particular forest will be saved, the surroundings will be filled with wind power plants, emitting sound, lights, microplastics, and PFAS, which along with the surrounding infrastructure will render reindeer husbandry here impossible. (Photo: May-Britt Öhman)

(Svenska Kraftnät, n.d.). There has been a major increase of traffic and several accidents and incidents with trucks hitting reindeer herders' vehicles. During one single week there were as many as four accidents. Luckily, there have been no personal injuries so far. The roads are narrow, and the trucks pass at three-minute intervals. Reindeer herders report that the truck drivers, in particular those from other European countries, drive at high speeds, far beyond the allowed speed limit. When an accident happened in early 2025, the truck driver stated that they drive fast as they do not think anyone lives there and are not well informed about the legal protection of reindeer. Preliminary insights also indicate that not all reindeer killed in traffic accidents are reported, which is a violation of the law (Reindeer herder, personal communication, February 15, 2025) (Fig. 9.6).

Concluding Remarks

Indigenous peoples have historically been negatively impacted due to settler colonialism, and this situation continues. Today, the exploitation of natural resources that takes place on Indigenous territory is claimed to be justified by societal demands of so-called clean and sustainable wind energy. However, violating Indigenous and human rights, and the legislation that is in place to protect reindeer herding, as well as constructing ecologically disturbing wind power plants are perpetuating destructive settler colonization rather than creating a transition towards sustainability. The Hällberget case is just one of several examples of how Indigenous Sámi rights are being violated today, in the name of a green transition. In the Norwegian Fosen case, the Norwegian Supreme Court ruled that Sámi rights were violated by the establishment of wind power. The court ruled that the political decision to allow construction was invalid, based on Article 27 of the International Covenant on Civil and Political Rights (UNGA, 1966). However, the Norwegian Government did not act according to the verdict, which led to a major conflict. This was followed by an agreement on monetary compensation and the recognition of the need for new land for winter grazing, and also strengthening of the reindeer herders' rights with a veto right on any future wind farm licences (Mósesdóttir, 2024).

Another case, in Sweden, is the Markbygden wind power industrial area, which when finalized will be Europe's largest land-based wind power industry, with 1100 wind power plants, located within the winter grazing lands of the Östra Kikkejaur Forest Sámi community. The Sámi community tried to oppose this, but it was impossible, and rather than fighting the government's 2009 decision in court, which is financially risky and also very difficult, Östra Kikkejaur opted to seek the best possible compensation agreement. Here, the winter lands have become completely unavailable and instead of grazing—finding their own food—the reindeer have to be fed with fodder during winter. This method of feeding entails extra work and transport using vehicles powered by fossil fuels on an everyday basis (Renkens, 2024). Furthermore, feeding reindeer with fodder may have negative impacts on their ability to find their own food and on their health, and may also have consequences for the biological diversity that natural reindeer grazing promotes (Rautiainen, 2024; Ramirez et al., 2024).

An important aspect of the situation with wind power and reindeer herding is that reindeer herders face a double punishment in the name of the green transition. Reindeer herders are already experiencing the consequences of climate change (Axelsson-Linkowski et al., 2020). There is a need for drastic societal changes, to reduce environmental destruction. However, when energy projects are established on Indigenous territories, against the Indigenous peoples' free, prior, and informed consent, they are punished at the expense of their rights and land use, the lands, and the waters. As mentioned, they are also negatively affected by the ungreen design of today's wind power (Helgesen et al., 2024; Renkens, 2024; Öhman et al., 2021).

For an actual sustainable transition—social, economic, and ecological—settler colonial behaviour towards Indigenous peoples needs to end. Sámi reindeer herder communities have to be included in the decision making. They should not just be offered monetary compensation when damage is already done. These so-called compensations do not cover actual expenses, and do not protect from life-threatening situations during construction periods, and provide a false image that the conflict is solved. The relationship with the lands and the reindeer is an inherent part of Sámi culture, livelihood, and wellbeing that monetary compensations cannot replace. This relationship needs to be respected and protected, and the rights enshrined in Swedish legislation upheld.

The UN Agenda 2030 and the EU Green Deal need to consider minorities' and Indigenous peoples' rights, and efforts need to be made to enforce these values in any actions and policy making that are referred to as being in the name of sustainability and climate change mitigation.

Acknowledgements We thank all reindeer herders in the Várre-group for sharing their knowledge and stories with us, and all other knowledge keepers who have shared insights. Thanks to the editors, anonymous peer reviewers, and language editing support. Funding for our research and this publication has been granted through several research project funded by Formas Dnr 2023-00772, the Göran Gustafsson Foundation, Formas Dnr 2021-01723, Dnr 2023-01829, Vetenskapsrådet Dnr 2021-03080, and Ărramăt, funded by the Government of Canada's New Frontiers in Research Fund (NFRF), [NFRFT-2020-00188].

Ethical approvals granted from the Swedish Ethical Review Authority Dnr 2023-02775-01 and Dnr 2024-07932-01.

References

Alves Dias, P., Bobba, S., Carrara, S., & Plazzotta, B. (2020). The role of rare earth elements in wind energy and electric mobility, EUR 30488 EN. Publication Office of the European Union, Luxembourg, https://doi.org/10.2760/303258.

Andersson, H., Storlöpare, P., & Öhman, M-B. (2019). Winds of destruction: A documentary project on wind power and human security – Hällberget, Överkalix, Gällivare Forest Sámi Reindeer Herding area [Video] Dálkke. *Youtube*. https://www.youtube.com/watch?v=x7ZylkDXxe8&t=671s.

Andersson, K. (2023). *Vuovddesáme/Skogssamer i Flakaberg*. Vulkan.

Andersson, K. (2024). *Unna Sájvva: en skändad samisk offerplats*. Vulkan.

Axelsson-Linkowski, W., Fjellström, A. M., Sandström, C., Westin, A., Östlund, L., & Moen, J. (2020). Shifting strategies between generations in Sami reindeer husbandry: The challenges of maintaining traditions while adapting to a changing context. *Human Ecology, 48*, 481–490. https://doi.org/10.1007/s10745-020-00171-3

Bennerhag, C., & Söderholm, K. (2024). Advanced ancient steelmaking across the Arctic European landscape. *Open Archaeology*, 10. https://doi.org/10.1515/opar-2024-0017.

Bergman, I. (2021). History, hearths and contested landscapes: Sámi land use east of the Lapland border of Sweden. *META – Historiskarkeologisk Tidskrift*, 109–120. https://doi.org/10.59008/meta.vi.10915.

Bjørklund, I. (2013). Domestication, reindeer husbandry and the development of Sámi Pastoralism. *Acta Borealia, 30*(2), 174–189. https://doi.org/10.1080/08003831.2013.847676

Cambou, D., Sandström, P., Skarin, A., & Borg, E. (2021). Reindeer husbandry vs. wind energy: Analysis of the Pauträsk and Norrbäck court decisions in Sweden. In M. Tennberg, E. G. Broderstad, & H.-K. Hernes (Eds.), *Indigenous peoples, natural resources and governance: Agencies and interactions* (Routledge Research in Polar Regions) (pp. 39–58). Routledge. https://doi.org/10.4324/9781003131274-3

Charmaz, K. (2017). The power of constructivist grounded theory for critical inquiry. *Qualitative Inquiry, 23*(1), 34–45. https://doi.org/10.1177/1077800416657105

Council of the EU. (2024). *European Green Deal.* European Council. https://www.consilium.europa.eu/en/policies/green-deal/.

Egelkraut, D., Aronsson, K. Å., Allard, A., Åkerholm, M., Stark, S., & Olofsson, J. (2018). Multiple feedbacks contribute to a centennial legacy of reindeer on Tundra vegetation. *Ecosystems, 21*, 1545–1563. https://doi.org/10.1007/s10021-018-0239-z

Eidlitz Kuoljok, K. (1969). *Food and emergency food in the circumpolar area* [Doctoral Dissertation]. Uppsala University.

Energimyndigheten. (2024). *Wind power statistics 2023.* https://pxexternal.energimyndigheten.se/pxweb/en/Energimyndighetens_statistikdatabas/Energimyndighetens_statistikdatabas__Officiell_energistatistik__Vindkraftsstatistik/EN0105_5.px/.

Eriksson, E. (2021). *Vinterbetesområdets förändring för Flakabergsgruppen i Gällivare skogssameby över två renskötande generationer* [Bachelor thesis] Swedish University of Agricultural Sciences, Umeå. Epsilon Archive. https://stud.epsilon.slu.se/17468/.

European Chemicals Agency. (n.d.). *Per- and polyfluoroalkyl substances (PFAS).* https://echa.europa.eu/hot-topics/perfluoroalkyl-chemicals-pfas.

European Commission. (2019). *The European Green Deal.* https://eur-lex.europa.eu/resource.html?uri=cellar:b828d165-1c22-11ea-8c1f-01aa75ed71a1.0002.02/DOC_1&format=PDF.

European Union. (2024). *Consolidated version of the Treaty on European Union.* http://data.europa.eu/eli/treaty/teu_2016/2024-09-01.

Eurostat. (2024). *Electricity from renewable sources up to 41% in 2022.* European Union. https://ec.europa.eu/eurostat/en/web/products-eurostat-news/w/ddn-20240221-1

Fernández-Llamazares, Á., Fa, J. E., Brockington, D., Brondízio, E. S., Cariño, J., Corbera, E., Farhan Ferrari, M., Kobei, D., Pernilla, V., Márquez, G. Y. H., Molnár, Z., Tugendhat, H., & Garnett, S. T. (2024). No basis for claim that 80% of biodiversity is found in Indigenous territories. *Nature, 633*(8028), 32–35. https://doi.org/10.1038/d41586-024-02811-w

Helgesen, I. S., Johannesen, A. B., Bostedt, G., & Sandorf, E. D. (2024). Climate change and reindeer herding – A bioeconomic model on the impact of climate change on harvesting profits for Saami reindeer herders in Norway and Sweden. *Ecological Economics, 223*, 108227. https://doi.org/10.1016/j.ecolecon.2024.108227

IEA. (2021). *The role of critical minerals in clean energy transitions*, IEA, Paris. https://www.iea.org/reports/the-role-of-critical-minerals-in-clean-energy-transitions.

Inga, B. (2008). *Traditional ecological knowledge among reindeer herders in northern Sweden.* [Licentiate Dissertation] Swedish University of Agricultural Sciences, Umeå. Epsilon Archive. https://pub.epsilon.slu.se/1947/1/inga_b_lic_2008_20090303.pdf.

Keating, D. (2023, June 15). Health vs climate? How an EU chemical ban could hinder the energy transition, *Energy Monitor.* https://www.energymonitor.ai/sectors/industry/health-vs-climate-pfas-how-an-eu-chemical-ban-could-hinder-the-energy-transition/?cf-view.

Labba, K. (2016). The legal organization of Sami reindeer herding and the role of the Siida. In K. Labba, C. Allard, & S. F. Skogvang (Eds.), *Indigenous rights in Scandinavia* (pp. 141–153). Routledge. https://doi.org/10.4324/9781315588353-13

Laframboise, L. (2023, September 12). The Sámi Limbo: Outlining nearly thirty years of EU-Sápmi relations. *The Arctic Institute*. https://www.thearcticinstitute.org/sami-limbo-outlining-nearly-thirty-years-eu-sapmi-relations/.

Lantto, P., & Mörkenstam, U. (2008). Sami rights and Sami challenges: The modernization process and the Swedish Sami movement, 1886–2006. *Scandinavian Journal of History, 33*(1), 26–51. https://doi.org/10.1080/03468750701431222

Lanyon-Garrido, C., Spik, S., Spik Skum, K., Öhman, M.-B., Helsdotter, E. C., & Tengö, M. (2024). A decolonial understanding of Sámi landscapes and human-nature relations in Sweden. In M. McEachrane & L. Faye (Eds.), *Decolonial Sweden* (pp. 39–58). Routledge. https://doi-org.ezproxy.its.uu.se/10.4324/9781003396611

Larsson, G. (2021). Forest Saami heritage and history Fjellström. In A. Salmi, M. Spangen, & T. Äikäs (Eds.), *Currents of Saami pasts: Recent advances in Saami archaeology* (pp. 121–148). The Archaeological Society of Finland.

Larsson, J., & Päiviö Sjaunja, E.-L. (2022). *Self-governance and Sami communities: Transitions in early modern natural resource management*. Springer Nature.

Laula, E. (1904). *Inför lif eller död?: sanningsord i de lappska förhållandena*. Svenska förlaget.

Lindvall, D. (2023). Why municipalities reject wind power: A study on municipal acceptance and rejection of wind power instalments in Sweden. *Energy Policy, 180*, 113664. https://doi.org/10.1016/j.enpol.2023.113664

Mathiesen, S. D. (2023). Reindeer husbandry in the circumpolar North. In S. D. Mathiesen, I. M. G. Eira, E. I. Turi, A. Oskal, M. Pogodaev, & M. Tonkopeeva (Eds.), *Reindeer husbandry* (pp. 1–13). Springer. https://doi.org/10.1007/978-3-031-17625-8_1

Mishnaevsky, L., Jr., Tempelis, A., Belahurau, Y., & Johansen, N. F.-J. (2024). Microplastics emission from eroding wind turbine blades: Preliminary estimations of volume. *Energies, 17*(24), 6260. https://doi.org/10.3390/en17246260

MMD. (2021). M 3771-19. Umeå Tingsrätt. https://www.domstol.se/umea-tingsratt/nyheter/2021/06/vasa-vind-ab-far-tillstand-till-avgransad-del-av-ansokt-vindkraftspark-i-overkalix-kommun/.

Mósesdóttir, L. (2024). Energy (in)justice in the green energy transition. The case of Fosen wind farms in Norway. *Technology in Society, 77*, 102563. https://doi.org/10.1016/j.techsoc.2024.102563

Mumford, E. (2021). *Marginalized indigenous knowledge and contemporary Swedish Colonialism: The case of reindeer husbandry in Gällivare Forest Sámi Community* [Master thesis] Uppsala University. Diva portal. https://www.diva-portal.org/smash/record.jsf?pid=diva2:1564585.

Naturskyddsföreningen i Norrbotten. (2024, May 20). *Vasa Vind bryter mot domstolsbeslut*. https://norrbotten.naturskyddsforeningen.se/vasa-vind-bryter-mot-domstolsbeslut/.

Nordin Jonsson, Å. (2010). *Árbediehtu/Árbbedietho/Aerpimaahtoe: Sametingets policydokument för traditionell kunskap*. Sametinget. https://www.sametinget.se/4180.

Norstedt, G. (2018). A land of one's own: Sami resource use in Sweden's boreal landscape under autonomous governance [Doctoral Dissertation] Swedish University of Agricultural Sciences, Umeå. Epsilon Archive. https://pub.epsilon.slu.se/15451/1/norstedt_g_180502.pdf.

O'Brien, J. M. (2017). Tracing settler colonialism's eliminatory logic in "Traces of History.". *American Quarterly, 69*(2), 249–255.

Öhman, M.-B. (2021). The ski or the wheel?: Foregrounding Sámi technological innovation in the arctic region and challenging its invisibility in the history of humanity. In B. Hokowhitu, A. Moreton-Robinson, L. T. Smith, S. Larkin, & C. Andersen (Eds.), *The Routledge handbook of critical indigenous studies* (pp. 431–446). Routledge.

Öhman, M.-B. (2024). Morfars farmors syster, mamma och jag: Återtagande och synliggörande av lulesamisk och skogssamisk historia och identitet som antikolonial och antirasistisk praktik. In K. Krifors, H. Sältenberg, D. Mulinari, & A. Neergaard (Eds.), *Antirasismer och antirasister.* Kriterium. https://doi.org/10.60156/kriterium.59.p

Öhman, M.-B., Andersson, H., Helsdotter, E. C., & Storlöpare, P. (2021). *Ungreen Windpower: Sámi Indigenous and scientific perspectives on fossil dependent and environmentally destructive designs.* [Video] Dálkke. Youtube https://www.youtube.com/watch?v=vMq-Yah4RL0.

Össbo, Å. (2021). 'A constant reminder of what we had to forfeit': Swedish industrial colonialism and intergenerational effects on Sámi living conditions in the area of upper Stuor Julevädno. *International Journal of Critical Indigenous Studies, 14*(1), 17–32. https://doi.org/10.5204/ijcis.v14i1.1629

Ramirez, J. I., Sundqvist, M., Lindén, E., Björk, R. G., Forbes, B. C., Suominen, O., Tyler, T., Virtanen, R., & Olofsson, J. (2024). Reindeer grazing reduces climate-driven vegetation changes and shifts trophic interactions in the Fennoscandian tundra. *Oikos, 2024*(11). https://doi.org/10.1111/oik.10595

Rautiainen, H. (2024). *Effects of winter-feeding on reindeer's future ability to utilize natural pastures* [Doctoral Dissertation]. Swedish University of Agricultural Sciences, Umeå. Epsilon Archive. https://pub.epsilon.slu.se/32344/1/rautiainen-h-20231219.pdf.

Renkens, I. M. (2024). Experiencing the double blow of climate change and its mitigation: Perspectives from Sámi reindeer herders in Swedish Sápmi. *Journal of Political Ecology, 32*(1). https://doi.org/10.2458/jpe.5769.

Salman, S. (2015, April 28). *Överkalix kommun sa ja till vindkraft.* P4 Norrbotten. https://www.sverigesradio.se/artikel/6152482.

Salmi, A.-K., Äikäs, T., Fjellström, M., & Spangen, M. (2015). Animal offerings at the Sámi offering site of Unna Saiva: Changing religious practices and human-animal relationships. *Journal of Anthropological Archaeology, 40*, 10–22. https://doi.org/10.1016/j.jaa.2015.05.003

Sametinget. (2024). *Markanvändning.* https://www.sametinget.se/underlag.

SBKL. (2018). *Maria Katrina (Karin) Stenberg*, www.skbl.se/sv/artikel/MariaKatrinaKarinStenberg, Svenskt kvinnobiografiskt lexikon (by Anne Heith), 2024-12-10.

SFS 1971:437 Rennäringslag (Swedish Reindeer Husbandry Act).

SFS 1974:152 Regeringsformen (Instrument of Government, Swedish Constitution).

Skarin, A., Sandström, P., Niebuhr Brandao, B., Alam, M., & Adler, S. (2021). *Renar, renskötsel och vindkraft.* Rapport 7011. Naturvårdsverket. https://www.naturvardsverket.se/4ac136/globalassets/media/publikationer-pdf/7000/978-91-620-7011-3.pdf.

Stoessel, M., Moen, J., & Lindborg, R. (2022). Mapping cumulative pressures on the grazing lands of northern Fennoscandia. *Scientific Reports, 12*, 16044. https://doi.org/10.1038/s41598-022-20095-w

Storlöpare, P., Öhman, M.-B., Andersson, H., Eriksson, E., Eriksson, M., & Eriksson, T. (2024). Ungreen windpower: Dutch retirement funds destroying reindeer lands [Video] Dálkke. *Youtube*. https://www.youtube.com/watch?v=IJPKUIiyL2s.

Svenska kraftnät. (n.d.). *Aurora Line*. https://www.svk.se/utveckling-av-kraftsystemet/transmissionsnatet/transmissionsnatsprojekt/aurora-line/.

UNGA. (1966). *International Covenant on Civil and Political Rights*. United Nations General Assembly. https://www.ohchr.org/en/instruments-mechanisms/instruments/international-covenant-civil-and-political-rights.

UNGA. (2015). *Transforming our world: The 2030 Agenda for Sustainable Development*. United Nations General Assembly A/RES/70/1, October 21. https://www.refworld.org/legal/resolution/unga/2015/en/111816.

United Nations. (n.d.). *Fight racism- indigenous peoples*. https://www.un.org/en/fight-racism/vulnerable-groups/indigenous-peoples.

Vasa Vind. (2024). Hällberget. https://www.vasavind.se/wp-content/uploads/2024/03/Hs-Samradsunderlag-Hallberget_andringstillstand_20240321-allm.pdf.

Vasa Vind. (n.d.). *Our owner*. https://www.vasavind.se/en/om-oss/owners/.

Whyte, K. (2023). Indigenous environmental justice, renewable energy transition, and the infrastructure of sovereignty. In P. C. Rosier (Ed.), *Environmental justice in North America* (pp. 307–330). Routledge. https://doi.org/10.4324/9781003214380-13

Wilson, A., & Laing, M. (2019). Queering Indigenous education. In L. T. Smith, E. Tuck, & K. W. Yang (Eds.), *Indigenous and decolonizing studies in education: Mapping the long view* (pp. 219–230). Routledge. https://doi.org/10.4324/9780429505010-9

CHAPTER 10

Unsettling Hopes: Sustainability and the Politics of Future-Making

Marjo Lindroth and Heidi Sinevaara-Niskanen

Introduction

The chapter joins a critical line of scholarship that problematises the United Nations' 2030 Agenda and Sustainable Development Goals (SDGs) (e.g. Gabay & Ilcan, 2017; Krieg & Toivanen, 2021; Larsen et al., 2022). It contributes to unveiling the elements of injustice and inequality embedded in these goals by focusing on hope as an ingredient of future-making. Amidst the insecurities in the world today, hope seems to possess heightened value, figuring centrally in discussions spanning migration, security, the use of natural resources and climate change. The intensifying search for hope has also piqued the interest of scholars (e.g. Haro, 2010; Kleist & Thorsen, 2017; Lindroth & Sinevaara-Niskanen, 2022).

Undoubtedly, hope has been seen as a positive force in sustainability transitions. It has been perceived as the driver of youth action, more sustainable life choices and moves towards green(er) economies (e.g. Čeginskas et al., 2023; Ojala, 2017; Miller & Crane, 2018). The emphasis on hope and the inclination to construct hopeful messages has even become a societal norm (Head, 2016). While the SDGs hinge on hopes for a brighter future—aspirations that are very easy to agree with—we argue that the question of hope itself has remained under the radar.

M. Lindroth
Arctic Centre, University of Lapland, Rovaniemi, Finland
e-mail: marjo.lindroth@ulapland.fi

H. Sinevaara-Niskanen (✉)
Gender Studies, University of Lapland, Rovaniemi, Finland
e-mail: heidi.sinevaara-niskanen@ulapland.fi

R. Toivanen et al. (eds.), *Decolonizing the Sustainable Development Goals*,
https://doi.org/10.1007/978-3-032-14263-4_10

By building on the critical scholarship on the SDGs, social scientific problematisations of hope and the question of decolonisation, we dissect the links between hope, sustainability and the exercise of power. For the purposes of contextualising these discussions, we refer to the ongoing debates on resources and land use in northern Finland, Sweden and Norway, the traditional homelands of the Sámi. The political calls for sustainable development and, more recently, the emphasis on green transition have brought to the fore how understandings of land and resources collide. To a growing extent, the Sámi and their ways of using and relating to the land compete with, for example, wind- and water-based renewable energy production (e.g. Fjellheim, 2023; Össbo, 2023). These developments have made it pertinent to ask whose hopes—and thus what kinds of visions of the future—the SDGs serve.

The Sustainable Development Goals as a Global Setting for Hope

> This is no time to panic or lose hope. Instead, it is time to act, collectively, with a long-term vision and focus. (Junhua, 2023a, p. xv)

In the preface of the 2023 Global Sustainable Development Report "Times of Crises, Times of Change" the Under-Secretary-General for Economic and Social Affairs, Li Junhua, emphasises the importance of retaining faith in the future, of being hopeful. Assessing the progress on the United Nations Sustainable Development Goals, the scientific report laments that the world is lagging behind in its commitments and is, in fact, "more off track" than it was before (Times of Crises, Times of Change, 2023, p. 1). However, as Junhua's (2023a) words emphasise, retaining hope is crucial and it is not too late to act.

Adopted by United Nations member states in 2015, the 2030 Agenda for Sustainable Development and its 17 Sustainable Development Goals are considered the definitive global framework for addressing sustainability challenges (United Nations, 2024). The SDGs are the result of lengthy political negotiations and processes that can be traced back to Agenda 21 of the 1992 Rio Earth Summit (e.g. Sachs, 2017). The work leading up to the creation of the 2030 goals involved numerous aspirations that were finally consolidated into the 17 goals, which span "good health and well-being", "gender equality" and "climate action" (United Nations, 2024). Each is operationalised in the form of specific "targets", with these totaling 169. The scope of the goals is vast and the task of realising them is an ambitious one, as the UN assessment reports also demonstrate (e.g. The Sustainable Development Goals Report, 2023).

The criticism levelled at the SDGs points to their vagueness when it comes to realisability. As Wolfgang Sachs (2017, p. 2574) has put it, "...the SDGs are a complex matter, a comprehensive manual that is at once ponderous and visionary but that is non-binding, and lacks a sanction mechanism". Reflecting on the high-minded appearance of the SDGs, Sachs notes how the goals are politically appealing as they do not impose any binding requirements. He states how the SDGs are yet another link in the chain of "perennial noncommittal avowals" of governments (Sachs, 2017, p. 2574). This is a reflection of a more general "UNcritical" ethos—governments' comfort zone—where "win-win language" dominates and the need for structural transformation is evaded (Larsen et al., 2022, p. 20). Accordingly, the SDGs harbour a perception where everyone is allegedly to benefit from them.

On a more fundamental level, critical scholarship has pointed out the undemocratic and unjust origins and repercussions of the SDGs (e.g. Larsen et al., 2022; Salleh, 2016; Toivanen & Kmak, 2021). From the outset, the process of formulating the goals and devising their implementation excluded a large part of the world's population and their diverse knowledges. Therefore, the processes designed to implement the goals have been selective and lack the potential for deeper economic or political transformation. For example, the tools to realise the SDGs rely heavily on the Western liberal logic of the markets at the expense of those populations who have already been dispossessed by those markets (Larsen et al., 2022; Salleh, 2016). Indeed, the goals can end up increasing inequality, with one's success or "good life" being built on another's deprivation of those very things (Toivanen & Kmak, 2021).

Building on these critical insights, we argue that the question of hope—hiding in plain sight as it is in the SDGs—merits further investigation. As described by Peter Bille Larsen et al. (2022, p. 21), "...a new vocabulary of legitimacy" has been created through the SDGs. The SDGs create an illusion of transformation by "harbouring a hope or potential for change" that obscures power relations (Larsen et al., 2022, p. 21). While the hopes embedded in the SDGs appear equally shared and equally beneficial, they are not without power implications.

We argue that the very foundation of the SDGs rests on selective hopes—and thus particular visions of the future—that have a bearing already here and now. Unpacking the political dimensions of the development goals, Clive Gabay and Suzan Ilcan (2017, p. 338) observe that the SDGs have a particular affective aspect that produces "a sense of a world that is moving on a particular pathway and towards something that can produce a sustainable, goal-oriented, and 'better' world". Echoing the observations of Gabay and Ilcan, we go on to dissect the ways in which hope is operationalised to build a particular sense of the world and a "brighter" future. The future-driven nature of hope makes it serviceable indeed for these ends as hopeful messages are what we, as humans, want to hear. Yet, it is worth noting that hopes can be evoked without committing to their realisation.

The Politics of Hope

> This special edition of the Sustainable Development Goals Report reminds us that there is still much work to be done. [...] Yet, this report also provides a vision of hope by showcasing the progress the world has made so far and the potential for further advancements. (Junhua, 2023b, p. 3)

Whether the focus is individual attitudes and life choices or global development agendas, one sees an emphasis on the search for hope in sustainability discussions. Catastrophic predictions are now accompanied—if not fully replaced—by reassuring vocabularies of hope in order to encourage positive environmental action and faith in the problem-solving abilities of humanity (e.g. Čeginskas et al., 2023; Ojala, 2017; O'Neill & Nicholson-Cole, 2009; Raygorodetsky, 2017). The burgeoning calls for hope have become the norm to the extent that there is a social expectation to speak about sustainability issues in optimistic terms (e.g. Åhlvik et al., 2023). This exhortation to avoid conceiving doom and gloom scenarios has extended to the scholarly world as well. As Lesley Head (2016, p. 2) puts it, there "is deep cultural pressure in the West" to lean towards optimism even though scientific evidence might suggest otherwise.

The global search for hope has permeated a wide range of societal spheres spanning development, security, migration, justice and minority rights (e.g. Allen, 2021; Kleist & Thorsen, 2017; Lindroth & Sinevaara-Niskanen, 2022; Sejersen, 2020; Tängh Wrangel, 2018). Social scientific research has also noted the growing value of hope especially in the past two decades. Hope is not, however, a novel object of study as a social and philosophical question. For example, Ernst Bloch's (1986) work on hope and utopia from the mid-twentieth century has been influential for the study of social transformation (e.g. Browne, 2005; Zigon, 2009). In Blochian thinking, there is a profound faith in hope and its capacity to emancipate. In particular, Bloch's notion of hope as "Not-Yet"—a process of becoming and creating something new—has captured the imagination of contemporary scholars.

Not surprisingly, hope has evolved as a concept to explain and understand a range of social, political and legal struggles aiming for transformation (Hage, 2003; Lindroth & Sinevaara-Niskanen, 2024; Ottendörfer, 2019). Hope and its significance have been studied as an individual and collective affect (Anderson, 2016; Massumi, 2015), and its essence and sources have intrigued scholars who have studied people in devastating circumstances (Lear, 2008). Hope has also been analysed as a social and institutional expectation (Hage, 2003; Nuijten, 2004). In addition, an increasing body of studies has broached hope as a methodology for social change or critique (Eagleton, 2015; Miyazaki, 2004).

Studies on hope can be divided into two broad categories: those that perceive hope as an empowering force and those that highlight its restrictive facets. The former approach tends to dominate the understandings of hope, as is the case in sustainability discussions. In this line of study, hope is understood as a human inclination that stems from a desire for a better future—strongly echoing Bloch's ideas. It is precisely because hope is a human capacity that it can be summoned for transformative action. Studies on resistance and grassroots movements serve as cases in point (e.g. Miller & Crane, 2018; Nairn et al., 2022). Indeed, hope's political nature has been that of a force for liberation, especially for those pursuing a better life and more just conditions, such as migrants or victims of civil war (Kleist & Thorsen, 2017; Ottendörfer, 2019).

Indigenous peoples have been acknowledged particularly for their ability to harness hope amidst dispossession and, what is more, to offer hope for the rest of humanity as it now faces environmental degradation (Lear, 2008; Raygorodetsky, 2017). The ability of Indigenous peoples to retain their hope has even been regarded as radical (e.g. Haro, 2010; Lear, 2008). As Jonathan Lear (2008) notes, the radical hope of Indigenous peoples has enabled them—as peoples—to survive colonial intrusions and even the destruction of their way of life. Lear's historical account refers to the end of buffalo hunting and the removal of the Crow Nation from its lands. Against all odds, the Nation retained its hope. The more recent efforts of Indigenous peoples to regain their land and autonomy have also brought to the fore the role of hope (Dinerstein, 2015; Miyazaki, 2004).

Amidst the increasing calls for hope in the world, more and more studies draw attention to its multifaceted role in society. The darker dimensions of hope—those imposing restriction and stagnation—have inspired critical analyses (Hage, 2003, 2016; Lindroth & Sinevaara-Niskanen, 2022; Waldow et al., 2024). Ghassan Hage, examining the societal distribution of hope among different groups of people, notes how already marginalised populations are expected to constantly "dig for new forms of hope" (Hage, 2003, p. 21). In particular, he highlights a power dynamics more complex than the mere question of the distribution of hopes. He cites a need to recognise how "group A extracts from group B, for itself, the very resources that group B needs to hope" (Hage, 2016, p. 466). This "process of extraction and exploitation" of hope lays the very foundation of inequality in society (ibid.).

Such observations highlight that hope should not be perceived one-sidedly as a force for good or a liberating power that can be harnessed by marginalised groups. Hope can equally be put to work for exploitative purposes, and those in power can harness it to serve their own agendas. As Rebecca Coleman and Debra Ferreday (2010, p. 315) point out, the common understanding of hope as singularly positive conceals its more sinister workings. It is this equivocality, emblematic of hope, that has seen it used for contrary political ends (Lindroth & Sinevaara-Niskanen, 2021).

The equivocality of hope has also been recognised in sustainability studies. Although the scholarship mostly emphasises the ability of hope to fuel action, a critical strand of research has revealed that it may not be the most effective factor in motivating pro-environmental behaviour and solutions (Grund & Brock, 2019; Hornsey & Fielding, 2016; Åhlvik, 2023). Contrary to what is often intuitively assumed, hopeful messages do not encourage action any more than pessimistic ones do. In some cases, they may even have the opposite effect, producing inaction (Hornsey & Fielding, 2016). Indeed, the presence of future-driven hope in sustainability discussions has been shown to slow down and even hinder debate on concrete actions (Åhlvik, 2023). Interestingly, other feelings, such as grief, may actually prove to be drivers of a more effective politics if they are acknowledged and not bypassed by calls for optimism (Head, 2016).

The future-driven parlance fostering hope features markedly in the SDGs. While the aim of such messages may well be encouragement, this does not make the calls for hope any less political. As critical research on the societal role of hope has noted, politics operates in and through hope by encouraging some and muting other hopes (Haro, 2010; Nuijten, 2004). What is the political role of hope in SDGs in directing development? Whose hopes for the future are given the decisive role? It is worth bearing in mind that the ethos of hope that prevails in the SDGs hinges not only on the explicit token—the word itself—but also on attendant implications that something better is to come.

(De)Colonising the Future

> We are at a moment of truth and reckoning. But together, we can make this a moment of hope. I urge all Member States to make 2023 the moment when we jump-start progress on the SDGs, to create a more peaceful and prosperous future for all. (Guterres, 2023, p. 2)

As Secretary-General of the United Nations António Guterres states in the annual report assessing the progress of the SDGs, the goals will (allegedly) enable a *prosperous future for all.* Guterres' words on *truth and reckoning* are also highly interesting given the history of the goals being built on a future-looking ethos that has seen, among other things, the root causes of the global state of affairs left unaddressed. One of the significant questions neglected in the SDGs is colonialism and its role in the production of inequality both in the past and the present (e.g. Larsen et al., 2022). Clearly, then, the SDGs are in need of decolonisation. But what would it mean to decolonise these goals?

Here we turn to scholarship on colonialism, settler colonial studies in particular, in order to understand the complexities and workings of contemporary colonialism (e.g. Coulthard, 2014; Tuck & Yang, 2012; Veracini, 2015). As this scholarship highlights, settler colonialism is characterised by settlers coming "with the intention of making a new home on the land" (Tuck & Yang, 2012, p. 5). It is this interest in land that sets it apart from other types of colonialism (Veracini, 2011, 2023). Lorenzo Veracini (2011, p. 4) has observed that while colonialism in general "wants itself ongoing", settler colonialism "wants itself terminated". In other words, settler colonialism strives towards a situation where land is no longer an issue or something to struggle over. This desire on the part of settler colonial states to have things "settled" has given rise to practices such as assimilation and displacement, aimed at making Indigenous peoples "go away" (Veracini, 2011, pp. 2–3). Summarising this dynamic of erasure, Patrick Wolfe (2006, p. 388) notes how "[w]hatever settlers may say [...] the primary motive for elimination is not race (or religion, ethnicity, grade of civilization, etc.) but access to territory".

Despite such a settler colonial agenda, there are a number of ongoing and completed land rights negotiations between Indigenous peoples and states that, to a lesser or greater degree, give some autonomy to Indigenous peoples. However, the processes or their results often end up falling short of what the Indigenous nation in question was seeking (e.g. Cameron, 2015; Samson & Cassell, 2013). These land struggles are testament to the continuous presence of the settler colonial project. As Veracini (2011, p. 4) puts it, the very existence and resistance of Indigenous peoples demonstrates the incompleteness of this project.

In settler colonial conditions, decolonisation thus needs to address the question of land. Eve Tuck and Wayne Yang (2012) go as far as to note that decolonisation is exclusively a matter of land and perceptions of it. In their view, decolonisation "must involve the repatriation of land simultaneous to the recognition of how land and relations to land have always already been differently understood and enacted" (Tuck & Yang, 2012, p. 7). Such a take on decolonisation, as Tuck and Yang point out, would be very demanding on society because it "implicates and unsettles everyone" (ibid., p. 7). The perception that decolonisation always requires addressing the question of land entails radical consequences in the context of sustainability goals. The SDGs and their vision of a sustainable future reflect certain understandings and ways of relating to the land. Whose hopes for and relations to the land do the SDGs represent and further?

It is already evident that contradictions have arisen in the wake of the efforts to put the SDGs into practice. Environmentally destructive resource extraction—from tar sands to pipeline projects—has been based on a Western understanding of land (e.g. Guerrieri, 2019; Simpson, 2019). This perception has deemed rural areas, the Arctic being a case in point, to be areas that are not "properly" utilised or populated and are thus free to be exploited. More recently, this attitude might be seen as somewhat tempered by the offerings of compensation and redress to those suffering the impacts. However, the ways in which the green transition, among other sustainability-driven developments, approaches the use of land and the commons do not constitute a break with the pressures brought to bear by Western and colonial practices (e.g. Fjellheim, 2023; Haller, 2019; Lassila, 2024a). Scholarship focusing on the global South has already demonstrated the connections between sustainability policies and processes of land grabbing (Ashukem, 2020; Haller, 2019). The emphasis on the green transition and energy production has exacerbated existing land use conflicts and spawned new ones, the Arctic again being no exception.

Hopes in Land and Resources: Nordic Debates

> With momentum for decarbonization, the climate transition must be managed well to ensure that it is just, equitable and orderly. Vulnerable groups must be protected.... (Times of Crises, Times of Change, 2023, p. 24)

Despite the well-intentioned aims of the sustainability transition, its potentially disadvantageous impacts for some parts of the world's population have been acknowledged. Studies have shown how Indigenous peoples in particular are in a vulnerable position and disproportionally affected both by the changing environment and the new interventions made in the name of more sustainable development (e.g. Sánchez Contreras et al., 2023; Kramvig & Avango, 2021; Kuokkanen, 2023).

Using the notion "resource-desiring machine", Michael Simpson (2019) draws attention to the never-ending quest for resources of the settler colonial state. Studying the history and present of the extraction of the tar sand bitumen in Alberta, Canada, and its impact on Indigenous peoples, Simpson traces how a local resource is made into a national commodity. In his view, a natural resource is not "naturally a resource" but it is made into one through social and historical processes such as colonial dispossession (Simpson, 2019, p. 11). Imagining resources, that is, bringing them into being, entails a certain perception of the land and the people inhabiting it, including their relations to and understandings of that land (see also Haller, 2019).

In order to exemplify the tensions created by sustainability projects and the ways in which they imagine the land and resources, we discuss ongoing debates in northern Norway, Sweden and Finland. The developments we are discussing are impacting the region's Indigenous peoples, the Sámi, who have their own ways of relating to the land. While the Nordic countries have not typically been described as sites of settler colonialism, scholars have argued that the concept has relevance for studying the region. Rauna Kuokkanen (2023, p. 14) highlights that settler colonialism can help "to understand the Nordic states' sometimes oppositional policies and procedures" concerning the Sámi. In our view, these observations can contribute to seeing the obvious, yet subsumed, hegemonic hopes invested in the land.

One of the fundamental discrepancies between Indigenous perspectives and Western sustainability visions is how the land is perceived to be "utilisable". As scholarship has noted, state-led and market-driven understandings value the land as a quarry from which to extract resources (Ashukem, 2020; Haller, 2019). For those living on and off the land, the land is about relations with it. In this latter understanding, as Simpson (2019, p. 2) puts it, "…reciprocal and respectful relations with a lively world" are at the core, not the rationality that seeks to extract from the land. The understanding of land as something that yields profit entails a conception that the land—to be utilised—is devoid of meaningful action. Such an understanding puts those with a profoundly different relationship to the land in a position where they need to give proof that the land is, in fact, not unused (Kuokkanen, 2023; Simpson, 2019).

This controversy plays out in a topical case on wind energy in Norway. The state-licensed wind park on the Fosen peninsula—traditional reindeer-herding pastureland of the local Sámi—has sparked a lengthy legal process (see von der Pütten, this volume). In a historical decision in 2021, the Norwegian Supreme Court ruled that the wind park built in the reindeer-herding area was a violation of the cultural rights of the Sámi to practice reindeer herding. This ruling required substantial efforts on the part of the herders to prove that the wind turbines cause considerable harm. Eva Maria Fjellheim's (2023) courtroom ethnography on the legal process shows how the herders had to demonstrate the illegitimate nature of the wind park. While the wind park representatives argued that the herders have enough pastures and that the reindeer can coexist with the wind turbines, the Sámi had to prove how the wind park constituted a denial of a traditional practice of theirs and restricted them from "using and relating to the landscape according to their knowledges, values and norms" (Fjellheim, 2023, p. 147).

Overall, the processes of assessing environmental impacts of resource extraction have been criticised for their industry bias, failure to respect Indigenous and local knowledge, and emphasis on local adaptation. Delving into the processes of green mining in northern Finland, Maija Lassila (2024b, p. 626) has noted how impact assessments support "the naturalization of harm and replaceability". Such processes do not recognise, as Lassila (2024b, p. 636) puts it, "the kinds of knowledge and sense of place that make a place unique". Accordingly, opposing the profit-seeking understandings of the land places a substantial burden of proof on those who relate to the land differently. Åsa Össbo (2023), discussing energy developments in Sweden, captures the dynamic succinctly:

> While decision-makers and representatives of other land-use interests seem to have difficulties considering perspectives other than their own, reindeer herders are expected to understand the perspectives of the wind-power, mining or forest industries or the larger community of which they are a part. Simultaneously, they must safeguard their own perspectives and culture. (Össbo, 2023, p.116)

This leads to further controversy in how the land is perceived and sustainability envisioned. The view of land as a resource to be rendered utilisable entails an idea that land can be replaced and compensated. Critical scholarship has, however, pointed out that such replacement cannot be done without substantial negative consequences (Kuokkanen, 2023; Lassila, 2024b; López, 2021). For example, the idea of mitigating the loss of land by monetary compensation is problematic. As Kuokkanen (2023, p. 22) notes in the case of the Sámi, "[i]t is not possible to compensate for the lost pastures because money cannot buy new ones; there are no 'extra' lands or available pastures".

Another harmful idea is that when environmental protection is not possible at the site impacted by extraction, the harm caused to biodiversity can be compensated by protecting a different area. Citing a mining case in northern Finland, Lassila (2024b) notes how such an approach does not recognise the significance and uniqueness of the local environment and its role in providing a basis for local life, from livelihoods to various cultural practices. Protecting nature somewhere else is not commensurate with the local loss. Discussing mining in Kiruna in northern Sweden, Elisa Maria López (2021; see also

López, this volume) points out that though the entire city is being relocated to enable mining to expand, the measures taken do not treat everyone equally. For the region's Sámi community, such relocation is yet another event in their historical experiences of displacement. Significantly then, the meanings and impacts of displacement place a heavier burden on the Sámi compared to the majority population, the development of wind power and Sámi reindeer herding being concrete examples (Fjellheim, 2023; Kuokkanen, 2023).

In sum, hegemonic hopes for sustainability manifest themselves in the use of land and resources in a way that causes fundamental insecurity to some. The contemporary examples from the Nordic countries demonstrate the ceaseless demands that the Sámi are to adapt and endure the harm caused to the land in the name of sustainability. Having a historical perspective on the energy developments in Sweden, Össbo (2023) draws attention to the ways in which national "emergencies"—ranging from the need for more hydro power after World War II to that for green wind energy in the era of climate crisis—are dealt with at the expense of the Sámi. This past and present national approach to energy condemns the land that the Sámi live in to be "sacrifice zones" for the greater good of the society (see also Kuokkanen, 2023). Even where resource extraction is only being envisioned, it triggers uncertainty regarding future life in the affected area. This uncertainty is irreversible, as Lassila (2024b) points out in the case of mining, and continues to impact local peoples regardless of whether extraction plans materialise or not (see also Guerrieri, 2019). In other words, the uneven distribution of the impacts of sustainability endeavours is not only a matter of historical or present injustice but, perhaps even more so, one of cementing dispossession in the future.

In light of these Nordic examples, the hopes for green transition inscribed in the SDGs—for all their benevolent appearance—are not far removed from Simpson's (2019) notion of the "resource-desiring machine". The realisation of the hoped for sustainable future depends on a certain kind of utilisation and harnessing of resources. The extraction-based relation to land and resources is emblematic of Western and urban values and lifestyles. Those who locally experience the impacts of such sustainability hopes might have a very different view of land, resources and how to achieve a sustainable future.

Conclusions

> The Sustainable Development Goals are disappearing in the rear-view mirror, as is the hope and rights of current and future generations. [...] We can do better, and in moments of severe challenge, humanity has always come through. Now is another of those moments. (The Sustainable Development Goals Report, 2023, p. 4)

Assessing the progress made with respect to the SDGs, the 2023 annual report, despite its alarmist tone, emphasises that there is still hope for humanity. The rhetoric on a unified "humanity", however, blurs the question of responsibility, both in causing the current situation and devising actions towards a sustainable future. As Larsen et al. (2022, p. 21) point out, the SDGs phrase "the language of goals and responsibilities in shared terms, as if all humans are equally culpable for the crises". This language also makes the question of responsibility seem irrelevant when the future is envisioned. As we see it, this is fertile ground for the exercise of power in and through hope.

The Nordic resource debates vividly illustrate the paradox of sustainability efforts. The envisioned sustainable and green future has to play out somewhere, on some(one's) land. The developments in Norway, Sweden, Finland and elsewhere demonstrate how, whether the focus of the project is mining, hydropower or wind energy, the views and lives of the Indigenous and local people need to yield. Simpson's (2019, p. 1) observation of how "one group's natural resource can be another's dispossession" applies not only to what is commonly perceived as destructive use of land but also to what is thought of as "green" resource use. While previous scholarship has shown how sustainability aims facilitate land grabbing (Ashukem, 2020; Haller, 2019), this chapter has brought to the fore the so far neglected role of hope as a means of dispossession. Indeed, even hopes and hoping can be unequal and support existing hierarchical structures in the society. The very foundations of someone's hopes may be built on restricting or bypassing the hopes of others (Hage, 2016).

In sustainability discussions, where there is a heightened call for optimism, hope is loaded with power. The constant search for hope, all the while avoiding pessimistic scenarios which might well become reality, glosses over the harm done in the name of "good" hopes. By focusing on hopeful accounts, the vulnerability that some parts of the population already experience in the face of uncertainty will continue unabated into the future. This is the politics of hope inscribed in the SDGs, a politics whereby the prevailing optimism normalises the sacrifices that some have to make. The decolonisation of the SDGs will thus require unveiling the political nature and power dimensions of hopes in sustainability debates. Recognising the role of hopes enables us to see how futures are constructed and how such processes favour some futurities over others. Encouraged by this realisation, we ought to dissect the bases on which our Western sustainability hopes are built in order to understand the multiple ways in which they cement inequality and insecurity.

References

Allen, L. (2021). *A history of false hope. Investigative commissions in Palestine.* Stanford University Press.

Anderson, B. (2016). *Encountering affect: Capacities, apparatuses, conditions.* Routledge.

Ashukem, J.-C. N. (2020). The SDGs and the bio-economy: Fostering land-grabbing in Africa. *Review of African Political Economy, 47*(164), 275–290.

Bloch, E. (1986). *The principle of hope.* (Vol. 1). MIT Press.

Browne, C. (2005). Hope, critique, and utopia. *Critical Horizons, 6*(1), 63–86.
Cameron, E. (2015). *Far off metal river. Inuit lands, settler stories, and the making of the contemporary Arctic.* UBC Press.
Čeginskas, V. L. A., Mäki, M., & Sonck-Rautio, K. (2023). Editorial. Hope in times of crisis and transformation. *Ethnologia Fennica, 50*(2), 1–4.
Coleman, R., & Ferreday, D. (2010). Introduction: Hope and feminist theory. *Journal for Cultural Research, 14*(4), 313–321.
Coulthard, G. S. (2014). *Red skin, white masks. Rejecting the colonial politics of recognition.* University of Minnesota Press.
Dinerstein, A. C. (2015). *The politics of autonomy in Latin America. The art of organising hope.* Palgrave Macmillan.
Eagleton, T. (2015). *Hope without optimism.* University of Virginia Press.
Fjellheim, E. M. (2023). Wind energy in trial in Saepmie: Epistemic controversies and strategic ignorance in Norway's green energy transition. *Arctic Review on Law and Politics, 14*, 140–168.
Gabay, C., & Ilcan, S. (2017). Leaving no-one behind? The politics of destination in the 2030 Sustainable Development Goals. *Globalizations, 14*(3), 337–342.
Grund, J., & Brock, A. (2019). Why we should empty pandora's box to create a sustainable future: Hope, sustainability and its implications for education. *Sustainability, 11*(3), 893.
Guerrieri, V. (2019). The spatiality of hope: Mapping Canada's northwest energy frontier. *Globalizations, 16*(5), 678–694.
Guterres, A. (2023). Foreword. In *The sustainable development goals report 2023: Special edition. Towards a rescue plan for people and planet.* Retrieved March 15, 2024, from https://unstats.un.org/sdgs/report/2023/The-Sustainable-Development-Goals-Report-2023.pdf
Hage, G. (2003). *Against paranoid nationalism: Searching for hope in a shrinking society.* Pluto Press Australia.
Hage, G. (2016). Questions concerning a future-politics. *History and Anthropology, 27*(4), 465–467.
Haller, T. (2019). The different meanings of land in the age of neoliberalism: Theoretical reflections on commons and resilience grabbing from a social anthropological perspective. *Land, 8*, 1–22.
Haro, L. (2010). The affective politics of insurgent hope. In J. Horrigan & E. Wiltse (Eds.), *Hope against hope: Philosophies, cultures and politics of possibility and doubt* (pp. 183–206). Rodopi.
Head, L. (2016). *Hope and grief in the anthropocene. Re-conceptualising human-nature relations.* Routledge.
Hornsey, M. J., & Fielding, K. S. (2016). A cautionary note about messages of hope: Focusing on progress in reducing carbon emissions weakens mitigation motivation. *Global Environmental Change, 39*, 26–34.
Junhua, L. (2023a). Preface. In *Times of crisis, times of change: Science for accelerating transformations to sustainable development*, the 2023 Global Sustainable Development Report. Retrieved March 15, 2024, from https://sdgs.un.org/sites/default/files/2023-09/FINAL%20GSDR%202023-Digital%20-110923_1.pdf
Junhua, L. (2023b). Introduction. In *The sustainable development goals report 2023: Special edition. Towards a rescue plan for people and planet.* Retrieved March 15, 2024, from https://unstats.un.org/sdgs/report/2023/The-Sustainable-Development-Goals-Report-2023.pdf

Kleist, N., & Thorsen, D. (Eds.). (2017). *Hope and uncertainty in contemporary African migration*. Routledge.
Kramvig, B., & Avango, D. (2021). The multiple landscapes of Biedjovággi: Ontological conflicts on indigenous land. *Polar Record, 57*, e35.
Krieg, C. P., & Toivanen, R. (Eds.). (2021). *Situating sustainability: A handbook of contexts and concepts*. Helsinki University Press.
Kuokkanen, R. (2023). Is reindeer the new buffalo? Climate change, the green shift and manifest destiny in Sápmi. *Meridians: Feminism, Race, Transnationalism, 22*(1), 11–33.
Larsen, P. B., Haller, T., & Kothari, A. (2022). Sanctioning disciplined grabs (SDGs): From SDGs as green anti-politics machine to radical alternatives? *Geoforum, 131*, 20–26.
Lassila, M. M. (2024a). *Incommensurabilities of new Arctic extractive projects. Resistance to mining among Sámi, reindeer herders and local communities in Sápmi and Finnish Lapland*. Dissertationes Universitatis Helsingiensis 43/2024. University of Helsinki.
Lassila, M. M. (2024b). An irreplaceable place: Onto-epistemological contestation in the environmental impact assessment process of the green Anglo American Sakatti mine, Arctic Finland. *Society & Natural Resources, 37*(5), 623–643.
Lear, J. (2008). *Radical hope: Ethics in the face of cultural devastation*. Harvard University Press.
Lindroth, M., & Sinevaara-Niskanen, H. (2021). Politics of hope: Transformation or stagnation? In S. A. H. Hosseini, J. Goodman, S. C. Motta, & B. K. Gills (Eds.), *The Routledge handbook of transformative global studies* (pp. 230–242). Routledge.
Lindroth, M., & Sinevaara-Niskanen, H. (2022). *The colonial politics of hope: Critical junctures of indigenous-state relations*. Routledge.
Lindroth, M., & Sinevaara-Niskanen, H. (2024). The hope-colonialism nexus. In V. Waldow, P. Bargués, & D. Chandler (Eds.), *Hope in the anthropocene. Agency, governance and negation* (pp. 135–149). Edinburgh University Press.
López, E. M. (2021). *Transforming Kiruna: Producing space, society, and legacies of inequality in the Swedish ore fields* (Uppsala Studies in Cultural Anthropology) (Vol. 62). Acta Universitatis Upsaliensis.
Massumi, B. (2015). *Politics of affect*. Polity Press.
Miller, C., & Crane, J. (Eds.). (2018). *The Nature of hope: Grassroots organizing, environmental justice, and political change. Intersections in environmental injustice*. University Press of Colorado.
Miyazaki, H. (2004). *The method of hope: Anthropology, philosophy, and Fijian knowledge*. Stanford University Press.
Nairn, K., Sligo, J., Showden, C. R., Matthews, K. R., & Kidman, J. (2022). *Fierce hope: Youth Activism in Aotearoa*. Bridget Williams Books.
Nuijten, M. (2004). Between fear and fantasy: Governmentality and the working of power in Mexico. *Critique of Anthropology, 24*(2), 209–230.
O'Neill, S., & Nicholson-Cole, S. (2009). "Fear won't do it": Promoting positive engagement with climate change through visual and iconic representations. *Science Communication, 30*(3), 355–379.
Ojala, M. (2017). Hope and anticipation in education for a sustainable future. *Futures, 94*, 76–84.
Ottendörfer, E. (2019). Assessing the role of hope in processes of transitional justice: Mobilising and disciplining victims in Sierra Leone's Truth Commission and Reparations Programme. *Globalizations, 16*(5), 649–663.

Raygorodetsky, G. (2017). *The archipelago of hope. Wisdom and resilience from the edge of climate change*. Pegasus Books.

Sachs, W. (2017). The sustainable development goals and Laudato si': Varieties of post-development? *Third World Quarterly, 38*(12), 2573–2587.

Salleh, A. (2016). Climate, water, and livelihood skills: A post-development reading of the SDGs. *Globalizations, 13*(6), 952–959.

Samson, C., & Cassell, E. (2013). The long reach of frontier justice: Canadian land claims 'negotiation' strategies as human rights violations. *The International Journal of Human Rights, 17*(1), 35–55.

Sánchez Contreras, J., Matarán Ruiz, A., Campos-Celador, A., & Fjellheim, E. M. (2023). Energy colonialism: A category to analyse the corporate energy transition in the global south and north. *Land, 12*(1241), 1–15.

Sejersen, F. (2020). Brokers of hope: Extractive industries and the dynamics of future-making in post-colonial Greenland. *Polar Record, 56*(E22), 1–11.

Simpson, M. (2019). Resource desiring machines: The production of settler colonial space, violence, and the making of a resource in the Athabasca tar sands. *Political Geography, 74*, 102044.

The Sustainable Development Goals Report. (2023). *Special edition. Towards a rescue plan for people and planet*. Retrieved March 15, 2024, from https://unstats.un.org/sdgs/report/2023/The-Sustainable-Development-Goals-Report-2023.pdf

Times of Crises, Times of Change. (2023). *Times of crisis, times of change: Science for accelerating transformations to sustainable development*. The 2023 Global Sustainable Development Report. Retrieved March 15, 2024, from https://sdgs.un.org/sites/default/files/2023-09/FINAL%20GSDR%202023-Digital%20-110923_1.pdf

Toivanen, R., & Kmak, M. (2021). Exclusion and inequality. In C. P. Krieg & R. Toivanen (Eds.), *Situating sustainability: A handbook of contexts and concepts* (pp. 137–152). Helsinki University Press.

Tuck, E., & Yang, W. K. (2012). Decolonization is not a metaphor. *Decolonization: Indigeneity. Education & Society, 1*(1), 1–40.

Tängh Wrangel, C. (2018). *The use of hope: Biopolitics of security during the Obama Presidency* (Doctoral dissertation). University of Gothenburg. Retrieved March 15, 2024, from https://gupea.ub.gu.se/handle/2077/5470

United Nations. (2024). Department of Economic and Social Affairs. Sustainable Development. Retrieved March 15, 2024, from https://sdgs.un.org/goals

Veracini, L. (2011). Introducing: Settler Colonial Studies. *Settler Colonial Studies, 1*(1), 1–12.

Veracini, L. (2015). *The settler colonial present*. Palgrave Macmillan.

Veracini, L. (2023). *Colonialism: A global history*. Routledge.

Waldow, V., Bargués, P., & Chandler, D. (Eds.). (2024). *Hope in the anthropocene. Agency, governance and negation*. Edinburgh University Press.

Wolfe, P. (2006). Settler colonialism and the elimination of the native. *Journal of Genocide Research, 8*(4), 387–409.

Zigon, J. (2009). Hope dies last: Two aspects of hope in contemporary Moscow. *Anthropological Theory, 9*(3), 253–271.

Åhlvik, T. (2023). *The ambiguous nature of hope. How a discourse of hope shapes the communicative conditions for exploring challenges in the circular economy transition* (Acta Universitatis Agriculturae Sueciae 2023) (Vol. 93). Swedish University of Agricultural Sciences.

Åhlvik, T., Bergeå, H., Rödl, M. B., & Hallgren, L. (2023). When hope messages become the discursive norm: How repertoires of hope shape communicative capacity in conversations on the circular economy. *Journal of Environmental Planning and Management, 67*(9), 2017–2034.

Össbo, Å. (2023). Back to square one. Green sacrifice zones in Sápmi and Swedish policy responses to energy emergencies. *Arctic Review on Law and Politics, 14*, 112–134.

CHAPTER 11

Figurative Invisibility and Paradigmatic Blindness in Sustainable Development Discourse

Tracey Skillington

Introduction

What are known today as the traditional homelands of Arctic Indigenous communities extend across modern geo-political boundaries created by the colonial conquests of European and North American nation-states. The Inuit homeland, for instance, stretches from Chukotka in Russia, across Alaska and northern Canada, to Greenland, while members of the Athabascan (Athapaskan) language family reside in Alaska and the northern regions of Canada, and those of Sápmi, the homeland of the Sámi, extend across northern and central Norway, Sweden, Finland and the Kola Peninsula in Russia. What the Arctic's Indigenous communities share in common is a strong attachment to their ancestral lands and traditional practices (including subsistence harvesting, reindeer husbandry, herding, fishing, hunting and gathering) (Larsen et al., 2015). This close affinity with the natural environment is increasingly challenged by changes occurring as a consequence of climate change, including rising temperatures, wildfire devastation and a critical loss of biodiversity (Ford, 2021), which, in turn, impact the livelihoods, social networks and traditional cultural practices of Indigenous people (Bogdanova et al., 2021).

Those subject to some of the most dramatic climate-related changes are communities of the Russian Arctic where temperatures are rising four times faster than anywhere else due to the 'polar-dependent amplifying effect' (Akenteva et al., 2017; Edel'geriev & Romanovskaya, 2020). This, in turn, leads evapotranspiration to increase, winter snowpacks and permafrost to thaw earlier, and lighting activity to multiply (Holzworth et al., 2020), resulting in

T. Skillington (✉)
Department of Sociology & Criminology, University College Cork, Cork, Ireland
e-mail: t.skillington@ucc.ie

R. Toivanen et al. (eds.), *Decolonizing the Sustainable Development Goals*,
https://doi.org/10.1007/978-3-032-14263-4_11

an increased likelihood of wildfire (Davis et al., 2019; Chevychelov, 2019). In 2020, it was estimated that wildfires in Arctic Siberia emitted a total of 58 million tonnes of carbon, a record value since records began in 2003. The fear is that these emissions will accelerate climate change even further, as well as the frequency and severity of wildfires, creating a feedback loop of continuing destruction. Adding to these problems is black carbon contained in wildfire smoke which when deposited on snow or ice causes the surface to darken, making it more likely to absorb solar radiation, accelerate permafrost thaw, and increase the risk of further fires (Ford, 2022). Extreme temperatures during the summer of 2021, for instance, resulted in more than 6 million hectares (15 million acres) of forested lands being scorched by wildfires in Siberia, according to Russia's forestry agency, with smoke spreading out across the entire Arctic Circle (Stone, 2021).

Between 2000 and 2020, over 70% of all wildfires in Russia were reported to have occurred in the boreal forest belt of Siberia (Conard & Ponomarev, 2020; York et al., 2020; Kharuk et al., 2021). At high latitudes of Siberia the main source of fires (almost 90%) is lightning strikes in Larch forests or 'dry thunderstorms', with minimal precipitation occurring at high temperatures and low relative humidity (Kharuk et al., 2021; Ivanov & Ivanova, 2010). Lightning strike fires destroy the soil's insulating layer, causing microbes in the permafrost to come out of hibernation and break down organic carbon in the soil, releasing CO2 and methane.

Adding to these problems is the fact that Indigenous communities in the Russian Arctic have limited access to emergency services and poorly developed wildfire protection infrastructure, planning and services, intensifying critical health effects (Dyar et al., 2022) and economic losses (Canosa et al., 2023; Leskinen et al., 2020; Solovyeva & Kuklina, 2020).[1] These trends further increase sensitivity to the impacts of climate change by modifying human–environment connections (e.g., settlement patterns, land use, ecosystem services and habitats) and diminish what historically would have strengthened resilience (e.g., intergenerationally transmitted knowledge; see Naylor et al., 2020). Together, these impacts jeopardise prospects for the fulfilment of internationally recognised adaptation goals, including the goal to 'hold the increase in global average temperatures to well below 2oC' (Paris Agreement, Article 2) and 'reduce vulnerability to climate change' (Article 7).

We may note an important distinction here between acknowledging harm (the common practice of projecting knowledge of climate destruction onto its victims as 'the other') and the less common practice of taking the perspective and needs of those affected into account when making moral and political judgements as to an appropriate response (defining victims as members of 'our' community and acting accordingly). This is also reflected in the difference

[1] In 2021 in Yakutia, Siberia, PM2.5 levels, that is, small particles entering the bloodstream and damaging organs, were recorded as reaching airpocalypse levels, meaning 'immediate and heavy effects on everybody', following widespread wildfires in the region.

between 'cognising' (*Erkennen*) what destructive impacts climate change is having, for instance, on Indigenous ways of life and 'recognising' (*Anerkennen*) the moral imperative to act to alleviate suffering. When agents of justice deny the practical relevance of a duty to assist those disproportionately affected by global climate change, they willingly uphold an interpretive and implicatory denial (Cohen, 2001) of the relevance of international climate change agreements and human rights obligations more generally.[2]

What these circumstances reflect are global scenarios of 'non-ideal justice' (Rawls, 1971) where the universal relevance of institutional principles of sustainable development (e.g., 'sustainable economic growth', Western development paradigms) proves 'non-ideal' for many. For instance, when minimal efforts are made to put emissions reductions targets and human rights norms into practice in ways that amount to a statement of moral and political non-responsibility for the suffering of millions. This chapter assesses how more institutionally embedded forms of denial of the agency component of human rights obligations and non-ideal justice relations are articulated in international policy discourse on sustainable development. Rather than a literal denial of the facts of climate change, this is a denial of moral ethical duties to address wrongdoing and alleviate the suffering of growing numbers. The discussion will focus on how this denial contributes centrally to the 'figurative invisibility' of Indigenous communities in the Arctic, as change agents in addressing climate change challenges. Practices of denial and 'non recognition' (Honneth, 1996), arguably, sustain rather than undermine histories of oppression by promoting a 'looking past' (Honneth & Margalit, 2001) the concerns of Indigenous peoples and ensuring rights obligations to the same remain largely unallocated. In this way, sustainable development discourse carries discriminatory tendencies that need to be highlighted and critically challenged. The discussion below proposes the introduction of a more dialogic model of climate justice deliberation, for instance in the form of a climate change truth commission where practices of denial and non-recognition, as components of 'epistemic' Indigenous injustice, can be targeted and greater equity established in the distribution of climate change burdens and responsibilities.[3]

[2] As Cohen (2001, pp. 7–8) explains, 'interpretive denial' occurs when 'the raw facts' of some problem (e.g., climate change) and its advancement are not denied but 'a different meaning from what seems apparent to others' is given to its significance. In the case of 'implicatory denial', the impacts of climate change are, again, acknowledged but all attempts to radically transform social–ecological relations and shift attention away from Western models of development are resisted (Cohen, 2001, pp. 8–9). Agents of justice are those with the capacity and sometimes the commitment to exercise agency in justice-relevant ways. While all social actors are, potentially, agents of justice, not all are willing to bring about justice-relevant change.

[3] What Fricker describes as 'epistemic injustice' can be considered relevant to consistent efforts to silence alternative knowledge perspectives on nature that do not align with a dominant Western one, as well as the wronging of specific subjects in their capacity to be knowledge agents (Fricker, 2007).

The Social Invisibility of Indigenous Peoples' Climate Action Skills

A consistent failing of the prevailing justice order is its denial of the ways it reproduces institutional patterns of discrimination and value inequality (McNay, 2008) and exacerbates experiences of climate change disadvantage (Brugnach et al., 2014). The discussion below assesses the type of interpretive strategies used to deny these connections. It examines some of the framing strategies used by state actors to play down discrepancies arising between normative commitments to equality, responsibility, respect and care (e.g., collective Indigenous rights) and 'actually existing' relations of injustice (historically conditioned), affecting the life chances and adaptive capacities of peoples facing large-scale catastrophe. Denial here includes a 'looking past' the experiences and knowledge perspectives of Indigenous peoples or their representatives even in settings widely thought to be 'inclusive'.

Both the *Climate Change Synthesis Report: Summary for Policy Makers* (2023) produced by the Intergovernmental Panel on Climate Change and the UN Environment Programme *Spreading like Wildfire* (2022) identify a need for the international community to draw on a more diverse range of knowledge sources, including Indigenous knowledge expertise on stewardship of nature, to enhance understanding of persisting barriers in the way of resilience-building. Yet, the newest strategic plan of the United Nations Development Programme (2022–2025) lists primary partners in the drive to accelerate 'progress towards green, inclusive transitions' as (1) member states, (2) United Nations agencies, (3) international financial institutions, (4) civil society and non-governmental organisations, (5) academia and (6) the private sector. No reference is made to the role of Indigenous communities. Is this an oversight, we may ask, or a position consistent with non-recognition? At the state level, also, there is a clear reluctance to grant sufficient recognition to Indigenous climate action skills (knowledge of land, water, ice and fire stewardship). The costs of such a lack of recognition are borne heavily by Indigenous communities who traditionally practise controlled cultural burnings of land in early spring and late fall to burn off an overgrowth of flammable vegetation near villages and fire-prone forests where many Indigenous communities reside.

For the Gwich'in, a group of Athabaskan-speaking North American Indian tribes in eastern Alaska and Yukon, cultural burning practices involve burning meadows in the early spring when there's still snow on the ground. This is designed to protect the roots of plant systems, improve the nutritional value of local vegetation for the different species that depend on it, including muskrats, waterfowl and other migratory birds and, ultimately, prevent the spread of more uncontrollable fires through the large grasslands and the taiga forest areas. Cultural burnings thereby ensure wildland fires are containable. Burning off non-productive grasses encourages a wide variety of plant species to grow, attracting a more diverse range of animals and insects to the area, thereby increasing the carrying capacity of the land and, in turn, the food security of

Native people. Yet at the state policy level, there is a persisting reluctance to grant sufficient recognition to the legitimacy of these practices or acknowledge the wider value of Indigenous communities' knowledge skills (e.g., knowledge of local biodiversity, growth cycles, land, water and fire stewardship, the importance of controlled cultural burnings in the early spring and late fall as spiritual and social events, etc.), as reflected in the Canadian government's decision to uphold the criminalszation of traditional fire stewardship practices (criminalisation is a colonial legacy). As Amy Cardinal Christianson (quoted by Osgood, 2023), an Indigenous fire specialist with Parks Canada from Treaty 8 Mentis territory in Alberta, Canada, comments: 'Indigenous people have been removed from conversations about fire management. We want to be able to steward the land and use fire on our territories'.

The decision to continue to suppress Indigenous landscape management has resulted in an overgrowth of flammable vegetation near Indigenous lands and a disproportionately higher number of wildfires (see Edward Alexander, 2021).[4] The costs of this non-recognition of Indigenous communities' traditional practices of cultural burnings, as well as water and land stewardship, are borne heavily by Indigenous communities who represent just 5% of Canada's population overall but comprise 42% of wildfire evacuees in recent years, according to an audit published by Indigenous Services Canada in June 2023. The chief reason for this is the fact that most Indigenous Canadian communities are located near fire-prone forests, such as the boreal forests, where dried out vegetation and flammable grasses increase the risk of wildfires near villages.

Métis, Inuit and First Nations peoples in Canada are, on average, three times more likely to be displaced by uncontrolled wildfires than the national average (Binon, 2024). In August 2023, almost 70% of the population of the Northwest Territories, including Yellowknife, the capital city, were evacuated due to wildfires (Aljazeera, 2023). The suppression of traditional Indigenous land management practices together with the effects of rising temperatures endangers the lives of many Indigenous communities. Equally, poor standards of health care, disproportionately higher rates of poverty and geographical isolation further disadvantage Indigenous communities. Alaskan Natives, for instance, have the lowest life expectancy of any ethnic group in the US or Canada due to poor access to health care and high rates of poverty (Blue Bird

[4] In 1874, British Columbia became the first province to ban cultural burnings, with other provinces following suit in the 1900s. The 1911 Report of the Superintendent of Forestry announced a new law where a $100 fine or three months in jail would be imposed on anyone caught burning the land. The explicit aim of these new rules was to 'assimilate Indigenous peoples' into Western cultural practices and relationships with fire (Boutsalis, 2020). In more recent years, there have been some efforts in Canada and the US to engage with Indigenous communities on this issue in 'multi-stakeholder schemes' regarding land rights and self-determination. However, according to Christianson (quoted by Boutsalis, 2020), such efforts to date have been 'localised and siloed' with few Indigenous fire experts allocated leadership roles. Adding to the experience of moral injury and non-recognition is the rule that government agencies must be informed of all burnings on Indigenous lands. 'Some First Nations do not agree with having to ask permission or report what they are doing on their [own] territory' (Christianson, quoted by Boutsalis, 2020).

et al., 2015). Similarly, life expectancy in Inuit-inhabited areas of Canada is more than 12 years lower than the Canadian average, while life expectancy in Greenland is nine years lower compared to non-Indigenous peoples. Suicide mortality amongst Indigenous communities is also eight times higher in Greenland (Granheim et al., 2021).

Thus, while wildfires are increasingly understood as linked to climate change (National Oceanic and Atmospheric Association, 2023), they are also linked to climate change scenarios exacerbated by acontextual or ethno-underdeveloped sustainable development initiatives. That is, Westernised approaches offering a generalised perspective on sustainable development 'from nowhere' (Benhabib, 1992)—acontextual, acultural and ahistorical —that flourish chiefly through a non-engagement with Indigenous knowledge systems (Zohar & Newhouse, 2019). To be effective, sustainable development policies must emerge from reciprocal engagements with the cultural practices and knowledge perspectives of different communities in their relations with natural and social worlds (Navarrete & Zohar, 2021). An 'ethno-developed' framework of Arctic fire stewardship, for instance, would look at the central role of culture in shaping social responses to climate change-related wildfires in the region.

Top-down, science-led sustainable development target approaches can be self-defeating when enforced without sufficient consideration for already existing, historically tested practices of caring for nature, including those that characterise Indigenous ways of life (e.g., banning controlled cultural burnings of flammable woodland vegetation). However, they can also be a harmful expression of non-recognition of the value of Indigenous Nations' knowledge of nature and relations of care. Instead, priority is given to the pyro economies of global capitalism based on expanding heat, rising temperatures and accelerating rates of resource destruction. The need to control the unpredictability of nature has facilitated its opposite—a loss of control—and now fire rages irrepressibly. The dynamic of acceleration in this instance is no longer experienced as liberating, as a life-giving force, but one of harm. As Edward Alexander (2022), Co-Chair of Gwich'in Council International, observes: 'We have had a front row seat to an Arctic on fire – a precursor to a world of flame'.

For Indigenous communities, relationships with fire must be seen as part of a wider ecology of fire that respects its powers and wisdom (e.g., the home fire that provides warmth and guidance and the community fire that brings people together). Indigenous fire stewardship is about fostering biodiversity, not destroying it, promoting ease of travel, and continuing the long-established cultural practices of their people (such as the tradition of the community fire).

Three Forms of Non-Recognition

We may note three forms of recognition (Honneth, 1996) that are actively withheld from Indigenous communities (Young et al., 2020). First, a withholding of the conditions allowing for the realisation of Indigenous peoples' potentials as sustainable development advocates. Second, a non-recognition of

Indigenous communities' rights to equal treatment before the law—where Indigenous communities are structurally excluded from enjoying certain rights, including a collective right to self-determine their own culturally sensitive responses to the effects of climate change and minimise the effects of harmful pollution practices. Third, a non-recognition of the unique qualities of Indigenous peoples' historical relations with nature and contributions to its flourishing (what McNay (2008) refers to as 'status subordination'). The example of Indigenous communities highlighted here draws attention to the intricate work involved in preserving relations of non-recognition, stereotyping, denial and cultural imperialism in climate change and sustainable development discourse. This work involves an active maintenance of conditions that sustain the impossibility of an identity and a culture that has no place in Western definitions of truth and relations with nature, as they are defined in various decision-making settings. The terms of official deliberation and policy-making are deliberately limited to dominant Western paradigms of reasoning.

One more prominent and dangerous aspect of this reasoning is presentism, where a distinct bias is displayed towards the present, or the 'here and now', as the most 'relevant' context of justice. Presentism is displayed, for example, in the preference for short-term electoral cycles or policy goals, as well as a more general limited degree of historical reflexivity in political consciousness. Presentism shapes ontological ways of being in the world where long-term trajectories on environmental risk are largely ignored in favour of 'myopic' perspectives that limit responsiveness to societal problems to the immediate present (Skillington, 2019). As long as justice is defined in these terms, what is right for all states to do in terms of ongoing acquisitions of limited land, seabed and atmospheric resources will not be considered in terms of what duties are owed to the peoples of the Arctic and other climate-vulnerable regions (Skillington, 2012). Instead, short-term gain will continue to take precedence over communal loss and insufficient attention will be accorded to the way historically embedded patterns of interaction with land, air and oceanic environments threaten the safety of multiple communities. What we may call the dominant political interpretive framework on climate change is characterised by high levels of denial and non-recognition (Honneth, 1996), including a denial of the relevance of Indigenous environmental agency. Reoccurring experiences of non-recognition and practices of rendering Indigenous communities politically invisible may be seen as part of a wider strategy to screen out non-Western visions of SDGs.

With the climate change and sustainable development debate still largely circumscribed around industrial capitalist interests and Western viewpoints (e.g., categorical distinctions between economy, environment, society and culture that are not recognised by Indigenous communities, leading to misalignments between Indigenous worldviews and sustainable development aims), the tendency is for differences to be cancelled out (Navarrete & Zohar, 2021). Without any real engagement with Indigenous perspectives on nature's protection, the 'otherness' of their position is maintained through an emphasis on their distance from the Western norm. The danger is that international sustainable development goals and Indigenous worldviews will become gravely misaligned not because of a lack of commitment to environmental imperatives but a non-recognition of difference. The latter usually occurs inconspicuously via routine habits of inattention to Indigenous detail and cultural stereotyping. In the process, discrimination and value inequality are actively maintained. It would appear there are wider strategies of containment at work in what seem on the surface to be 'inclusive' political arrangements (e.g., UN Committee on the Elimination of Racial Discrimination; Permanent Forum on Indigenous Issues; Local Communities and Indigenous Peoples Platform), leading to a consistent lack of sufficient support for Indigenous peoples' plight.

Non-recognition is expressed in various forms, including through a type of 'paradigmatic blindness' to the unique qualities of Indigenous knowledge perspectives and environmental practices, where the terms of official deliberation and policy-making are deliberately limited to dominant paradigms of reasoning. The concept of paradigm blindness implies that individuals, groups and organisations are unwilling to consider alternative perspectives or accept any challenge to the core principles of Western ways of making sense of the world (Cortright, 2015). The dominant SD paradigm produces its own literary forms (discursive conventions, vocabulary, themes, etc.) and methodologies through which the cultural and political dimensions of SD debate are defined as 'given' (remain non-contestable) with easily quantifiable or physically measurable targets (e.g., tools for measuring SDG 'indicators' in accordance with an 'either/or' positivist approach; see Mair et al. (2017) for further discussion). This distinctly Western approach is designed to preserve the relevance of focusing on science, economy, ecology, social and culture as separate or distinct silos, as well as a model of 'development as growth' (where development is quantified in reductive, material terms). No space is given to visions of sustainability and wellbeing that could potentially replace this model of separation with one of interconnection and the layering of relations between nature and culture (Redvers et al., 2023). For Indigenous community leaders such as Edward Alexander, Arctic Wildfire Sharing Circle, Co-Chair of Gwich'in Council International, the spirit of cooperation is missing from wider SD debate. Sharing firefighters and equipment is not enough. Wider cooperation is needed to create an effective preventive strategy. While cooperation may be central to the directional systems and philosophies of Arctic Indigenous communities, it is not necessarily central to the logic of Western systems of rational autonomy.

Even so, the Gwich'in Council International (GCI) is currently working with the Arctic Council to explore ways in which the Arctic nations can coordinate their responses to the growing threat posed by wildfires more effectively. One project run by the Council's Conservation of Arctic Flora and Fauna (CAFF) group maps wildland fires and documents best practices to better improve responsiveness to signals of danger, while another, run by its Emergency Prevention, Preparedness, and Response (EPPR) working group, is evaluating the possibility of bilateral agreements between different states and the creation of a new circumpolar agreement on rapid wildfire responses. In 2022, the Arctic Council's Working Groups Conservation of Arctic Flora and Fauna (CAFF) and Emergency Prevention, Preparedness, and Response (EPPR), in collaboration with Permanent Participant Gwich'in Council International (GCI), organised the Arctic Wildland Fire Sharing Circle event to bring Indigenous community representatives, project leaders and interested parties together for two days to consider such proposals and reflect further on the challenges they face as a community.

The necessity of such cooperation is driven home by the intensifying nature of risk. Wildfires and related fatalities are increasing in intensity across all Arctic regions in ways that point to the limited relevance of geo-political boundaries. In June 2023, for example, wildfires in Canada extended into the central metropolitan area of New York State, as well as parts of Pennsylvania and New Jersey, producing 'hazardous levels' of air pollution (US EPA Air Quality Index, 2023). Regardless of their geographic origins, wildfires, temperatures rises and greenhouse gas emissions have an impact everywhere, causing small-scale and universal ecological fates to continuously collide. In this way, climate risks are inherently cosmopolitan in the way they necessitate cooperation and the application of a more common earth reasoning (Skillington, 2017, 2018, 2019). Beyond such initiatives, there is little evidence in the wider geo-political context of the type of 'enlarged thinking' far-reaching cooperation requires.[5] Real dialogue would allow sufficient space for diverging perspectives to be heard and critically appraised. Instead, Western perspectives on SDGs (a science-led, a top-down approach focused on measuring quantifiable outcomes) are used as a regulative ideal, restricting the boundaries of what are considered 'relevant' contributions to sustainable development policy.

Conclusion

Enabling and protecting the ongoing 'ethno-development' of Indigenous communities in the Arctic requires that sustainable development goals be articulated from 'within the framework of larger society but on their own premises' (Buvollen, 1989, p. 130; Navarrate & Zohar, 2021). In other words, it requires the application of a more situated perspective on SDGs. For instance, plans for the realisation of Goals 15 (life on land) and 16 (peace, justice and strong institutions) should be worked through deliberatively and the continuity or discontinuity of certain cultural practices explored from the perspective of all affected parties (United Nations Environment Programme, 2022). To address the current reality of non-recognition or distorted communication of Indigenous communities' plight, it is necessary to devise a more inclusive framework of deliberation, regulation, accountability and rapid response to major risks. Devising such a framework, however, first requires that more institutionally

[5] Seyla Benhabib (1992) borrows the concept of 'enlarged thinking' from Kant (1952), Section 49 of *The Critique of Judgement*.

embedded problems of non-recognition and denial be addressed to ensure that the facts of inequality and current deformities in the representation of historical and contemporary truths are resolved satisfactorily (Skillington, 2017, p. 235). Normative principles of climate justice, rights and responsibilities must be situated more firmly within relational frameworks that speak to differences arising between peoples and their experiences of historical, geographical, generational and socio-cultural loss (Skillington, 2018, 2019, 2023). Arguably, such deliberative procedures will not succeed if they are not truly 'relational' or based on a genuine reciprocity of perspectives that ensures issues of non-recognition and wrongdoing are dealt with effectively (along with top-down approaches to the implementation of sustainable development goals, the ongoing mismanagement of dwindling energy, marine and land resources in a circumpolar and global context, poorly coordinated and under-financed resilience-building initiatives to help communities better manage the impacts of wildfire, large-scale loss of ice, livelihoods and traditional ways of life, etc.).

Confronting wrongdoing or guilt for historical and contemporary wrongs imposed on Indigenous communities means addressing the ways in which interpretive and implicatory denial continue to be used to delimit debate on issues of responsibility and decision-making authority. What is not considered adequately in international policy discourse is how current climate scenarios have been created by histories of oppression. Denial is strategically used to avoid a focus on those actors who, historically, have been the recipients of the benefits of material gains from commercial expansions in the Arctic and have, on that basis, accumulated duties of responsibility to its peoples, those for whom opportunities to adapt to the challenges posed by rising temperatures, wildfires, melting ice and other climate-related events have been curtailed by the activities of beneficiaries. Justice, therefore, requires that historical and contemporary wrongdoing which threatens the long-term survival of the Arctic's Indigenous communities is challenged. One option is to establish a climate change truth commission to bring much needed clarity to bear on the nature of wrongdoing, how it is experienced, and by whom, as well as ways in which corrective action can be initiated and minimum standards of democratic inclusion, accountability, recognition, responsibility and care implemented.[6] In

[6] Usually, corrective justice is understood as requiring a judgement of liability that 'rectifies the injustice inflicted by one party on another' (Weinrib, 2002, p. 349). Law, for instance, responds to such a judgement and establishes equality between parties by depriving one of the gains they accrued through their exploitation of the other and restoring it to the latter. In the case of a truth commission, corrective justice may also require a comprehensive revision of the narrative of history, in this instance, the history of the Arctic, to include the voices and testimonies of those Indigenous peoples formerly excluded, as well as a series of comprehensive, socially grounded measures to address wrongdoing, including new teaching and learning programmes, skills training, new sustainable development implementation and measurement tools, legally backed decision-making procedures, an SD policy review forum, as well as a new court of climate justice to address ongoing violations of the rights of Indigenous peoples. The results of any effective truth commission for Indigenous climate justice cannot be confined to a critical report but must be a catalyst for real social change and collective societal learning (one of the main criticisms of Norway's Truth

addressing wrongdoing, a climate change truth commission for the Arctic can help initiate stronger forms of social and political cooperation to counteract the social invisibility of the Indigenous subject and address the paradigmatic blindness displayed towards non-Western knowledge perspectives and cultural practices (what Honneth (1996) refers to as 'distorted recognition'). Such moves could potentially offer much needed clarity on how climate change is experienced, understood and responded to by the peoples of the Arctic. In other words, a deliberatively grounded truth commission could facilitate a move away from purely scientific approaches to the realisation of the SDGs to those that place communities, histories, livelihoods, testimonies of wrongdoing and corrective action genuinely at the centre of a more ethno-centred sustainable development framework.

References

Akenteva, E. M., Alexandrov, E. I., Alekseev, G. V., Anisimov, O. A., Balonishnikova, Z. A., Bulygina, O. N., Georgievsky, V. Y., Dokukin, M. D., Efimov, S. V., & Ivanov, N. E. (2017). *Report on climate risks in the territory of the Russian Federation.* Rosshydromet.

Alexander, E. (2021). Update - Circumpolar wildland fire project. Gwich'in Council. https://gwichincouncil.com/sites/default/files/2021%20EPPR%20June%20Circumpolar%20Fire%20update%20GCI.pdf

Alexander, E. (2022). Arctic wildland fire sharing circle. https://vimeo.com/showcase/9321788/video/683219228

Aljazeera. (2023). Huge wildfire forces evacuation of Canada's Northwest Territories capital. https://www.aljazeera.com/news/2023/8/17/huge-wildfire-forces-evacuation-of-canadas-northwest-territories-capital

Benhabib, S. (1992). *Situating the Self.* Polity Press.

Binon, E. (2024). *Canada's indigenous peoples increasingly at risk of disaster displacement.* Internal Displacement Monitoring Centre (IDMC). https://www.preventionweb.net/news/canadas-indigenous-peoples-increasingly-risk-disaster-displacement

Blue Bird, J. V., Peercy, M., Branam, D., Saunkeah, B., Wharton, D., Winkleby, M., et al. (2015). Beyond health equity: Achieving wellness within American Indian and Alaska Native communities. *American Journal of Public Health, 105*(3), S376–S379.

Bogdanova, E., Andronov, S., Soromotin, A., Detter, G., Sizov, O., Hossain, K., Raheem, D., & Lobanov, A. (2021). The Impact of Climate Change on the Food (In) security of the Siberian Indigenous Peoples in the Arctic: Environmental and lavri Risks. *Sustainability, 13*(5), 2561.

Boutsalis, K. (2020, September 20). Indigenous craft of cultural burning. *The Narwhal.* https://thenarwhal.ca/indigenous-cultural-burning/

and Reconciliation Commission investigating 'the Norwegianization policy and injustice against the Sámi and Kvens/Norwegian Finns' (2023) was its 'lack of teeth' in terms of the implementation of institutional change and support for a series of follow-up measures (Last, 2023)).

Brugnach, M., Craps, M., & Dewulf, A. (2014). Including indigenous peoples in climate change mitigation: Addressing issues of scale, knowledge and power. *Climate Change, 140*(1). https://doi.org/10.1007/s10584-014-1280-3

Buvollen, H. P. (1989). Regional autonomy in Nicaragua. A new approach to the indigenous question in Latin America. *Alternatives, 14*(1), 123–132.

Canosa, I. V., et al. (2023). Wildfire adaptation in the Russian Arctic: A systematic policy review. *Climate Risk Management, 39(10)*, 100481. https://doi.org/10.1016/j.crm.2023.100481

Chevychelov, A. P. (2019). Forest fires in yakutia and their influence on soil cover in the aspect of the forecast of immediated climate change. *Bull. NEFU. Earth Science Series, 1*(13), 55.

Cohen, S. (2001). *States of Denial.* Polity Press.

Conard, S. G., & Ponomarev, E. (2020). Fire in the North. https://www.iawfonline.org/article/fire-in-the-north-the-2020-siberian-fire-season/

Cortright, R. N. (2015). Applying research-based principles of learning to enhance educational practices. *Journal on Excellence in College Teaching, 26*(3), 5–22. https://doi.org/10.1080/08923641003773929

Davis, K. T., Dobrowski, S. Z., Higuera, P. E., Holden, Z. A., Veblen, T. T., Rother, M. T., Parks, S. A., Sala, A., & Maneta, M. P. (2019). Wildfires and climate change push low-elevation forests across a critical climate threshold for tree regeneration. *Proceedings of the National Academy of Sciences, 116*(13), 6193–6198.

Dyar, O. J., et al. (2022). Rainbows over the world's public health: Determinants of health models in the past, present and future. *Scandinavian Journal of Public Health, 50*(7), 1047–1058.

Edel'geriev, R. K., & Romanovskaya, A. A. (2020). New approaches to the adaptation to climate change: The Arctic Zone of Russia. *Russian Meteorology and Hydrology, 45*(5), 305–316.

Ford, A. (2022). *Wildfires as a shared Arctic variable.* Leverhulme Centre for Wildfires. Environment and Society. https://centreforwildfires.org/news/wildfires-as-a-shared-arctic-variable/

Ford, J. D. (2021). The rapidly changing Arctic and its societal implications. *WIREs Climate Change, 12*(6), e735.

Fricker, M. (2007). *Epistemic injustice: Power and the ethics of knowing.* Oxford University Press.

Granheim, I. P., Silviken, A., Viskum, C., Larsen, L., & Kvernmo, S. (2021). Socio-demographic, psychosocial and environmental factors associated with suicidal behaviour in Indigenous Sami and Greenlandic Inuit adolescents; the WBYG and NAAHS studies. *International Journal of Circumpolar Health, 80*(1), 1913939. https://doi.org/10.1080/22423982.2021.1913939

Holzworth, R. H., Brundell, J. B., McCarthy, M. P., Jacobson, A. R., Rodger, C. J., & Anderson, T. S. (2020). Lightning in the Arctic. *Geophysical Research Letters, 48*, e2020GL091366.

Honneth, A. (1996). *The struggle for recognition: The moral grammar of social conflicts.* John Wiley & Sons.

Honneth, A., & Margalit, A. (2001). Recognition. *Aristotelian Society Supplementary, 75*, 111–139.

Ivanov, V. A., & Ivanova, G. A. (2010). *Wildfires from lightning in forests of Siberia.* Nauka.

Kant, I. (1952). *The critique of judgment* (J. M. Meredith, Trans.). Clarendon Press.

Kharuk, V. I., Ponomarev, E. I., Ivanova, G. A., Dvinskaya, M. L., Coogan, S. C. P., & Flannigan, M. D. (2021). Wildfires in the Siberian taiga. *Ambio, 50*, 1–22.

Larsen, J. N., Fondahl, G., & Nordic Council of Ministers, Nordic Council of Ministers' Secretariat. (2015). *Arctic human development report: Regional processes and global linkages (AHDR-II)*. Nordisk Ministerrad.

Last, J. (2023, June 19). Norwegians get a bitter taste of their own history, *FP*. https://foreignpolicy.com/2023/06/19/norway-sami-indigenous-rights-truth-reconciliation-report/

Leskinen, P., Lindner, M., Verkerk, P. J., Nabuurs, G. J., Van Brusselen, J., Kulikova, E., Hassegawa, M., & Lerink, B. (Eds.). (2020). *Russian forests and climate change. What Science Can Tell Us 11*. European Forest Institute.

Mair, S., et al. (2017). A critical review of the role if indicators in implementing the sustainable development goals. In W. Leal (Ed.), *Handbook of sustainability science and research* (pp. 41–56). Springer.

McNay, L. (2008). The trouble with recognition: Subjectivity, suffering and agency. *Sociological Theory, 26*(3), 271–296.

Navarrete, M., & Zohar, A. (2021). Rethinking sustainable development by following Indigenous approaches to community wellbeing. *Tapuya: Latin American Science, Technology and Society, 4*, 1. https://doi.org/10.1080/25729861.2021.1946315

National Oceanic and Atmospheric Association (2023). Wildfire climate connection. https://www.noaa.gov/noaa-wildfire/wildfire-climate-connection.

Naylor, A., Ford, J., Pearce, T., & Van Alstine, J. (2020). Conceptualizing climate vulnerability in complex adaptive systems. *One Earth, 2*, 444–454.

Osgood, B. (2023). As Canada reels from wildfire, First Nations hope for larger role. *Aljazeera*. https://www.aljazeera.com/news/2023/7/5/as-canada-reels-from-wildfire-first-nations-hope-for-larger-role

Rawls, J. (1971). *A Theory of Justice*. Harvard University Press, Belknap Press.

Redvers, et al. (2023). Indigenous peoples: Traditional knowledges, climate change and health. *PLOS Glob. Public Health, 3*(10). https://doi.org/10.1371/journal.pgph.0002474

Skillington, T. (2012). Climate change and the human rights challenge: Extending justice beyond the borders of the nation state. *International Journal of Human Rights, 16*(8), 1196–1212.

Skillington, T. (2017). *Climate justice & human rights*. Palgrave.

Skillington, T. (2018). A deeper framework of cosmopolitan justice. In G. Delanty (Ed.), *Routledge international handbook of cosmopolitanism studies* (2nd ed.). Routledge.

Skillington, T. (2019). *Climate change and intergenerational justice*. Routledge.

Skillington, T. (2023). Thinking beyond the ecological present. *European Journal of Social Theory, 26*(2), 236–257.

Solovyeva, V., & Kuklina, V. (2020). Resilience in a changing world: Indigenous sharing networks in the Republic of Sakha (Yakutia). *Polar Record, 56*. https://doi.org/10.1017/S0032247420000406

Stone, M. (2021, August 19). Siberia's massive wildfires are unlocking extreme carbon pollution. *National Geographic*. https://www.nationalgeographic.com/environment/article/siberias-massive-wildfires-are-unlocking-extreme-carbon-pollution

United Nations Environment Programme. (2022). *Spreading like wildfire: The rising threat of extraordinary landscape fires*. https://www.unep.org/resources/report/

spreading-wildfire-rising-threat-extraordinary-landscapefires#:~:text=A%20new%20report%2C%20Spreading%20like,even%20in%20areas%20previously%20unaffected

US EPA Air Quality Index. (2023). https://www.epa.gov/outdoor-air-quality-data/air-quality-index-report

Weinrib, E. J. (2002). Corrective justice in a nutshell. *The University of Toronto Law Journal, 52*(4), 349–356.

York, A., Bhatt, U. S., Gargulinski, E., Grabinski, Z., Jain, P., Soja, A., Thoman, R. L., & Ziel, R. (2020). Wildland fire in high northern latitudes. NOAA Arctic Report Card. (2020). https://doi.org/10.25923/2gef-3964

Young, T. K., Broderstad, A. R., Sumarokov, Y. A., & Bjerregaard, P. (2020). Disparities amidst plenty: A health portrait of Indigenous peoples in circumpolar regions. *International Journal of Circumpolar Health, 79*(1), 1805254. https://doi.org/10.1080/22423982.2020.1805254

Zohar, A., & Newhouse, D. R. (2019). Educating for a sustainable world: Bringing together indigenous and western knowledges. In *Intellectual, scientific, and educational influences on sustainability research* (pp. 121–137). IGI Global.

CHAPTER 12

Empowerment of Indigenous Women: Insights from Nivkh Women on Sakhalin Island, Russia, and Kyrgyz *Koshok* Singers

Gulnara Aitpaeva and Vladislava Vladimirova

Introduction: Sustainable Development Goal 5 and Women's Equality

Women's equality serves an important role in the green transition according to the global Sustainable Development Goals (SDGs). This reinforces earlier development and human rights frames, such as the UN agenda and developmental goals, but extends them to the context of sustainability and climate change action. SDGs send an important message in a world where conservative politicians question women's liberties and capacities: environmental sustainability and green growth should not build on injustices, nor advance at the expense of societal groups, such as women, or Indigenous people. SDG5 elaborates how gender equality is to be implemented globally, drawing on values, imaginaries, and procedures from earlier feminist and developmental ideas and experiments.

SDG5 outlines several targets to help quantify gender equality and empowerment worldwide. Evaluation and comparison between states along common criteria and unified measurement units are thought to help their implementation. They give feedback to each state in relation to other states, and make it possible to identify the "leading states" in women's equality, such as Sweden,

G. Aitpaeva (✉)
Aigine Cultural Research Center, Bishkek, Kyrgyzstan

V. Vladimirova
Institute for Russian and Eurasian Studies/Department of Cultural Anthropology and Ethnology, Uppsala University, Uppsala, Sweden
e-mail: vladislava.vladimirova@ires.uu.se

R. Toivanen et al. (eds.), *Decolonizing the Sustainable Development Goals*,
https://doi.org/10.1007/978-3-032-14263-4_12

and those that lag behind. Comparative evaluations generate different statuses and symbolic capital and thus widen the gap between nations that are being thought of as "leaders" and as "underdeveloped." They also make the case for leaders to "transfer good practices" to the rest. The quantification of women's empowerment evolves from a presumption that there are particular models and qualities of life that are perceptible, discernable, and explicit enough to be measured. These features, or at least a significant proportion of them, are then seen as a necessary part of each woman's life in order for her to achieve well-being. Below, we show how SDG5 describes some of these requisite qualities of women's lives that are felt to lead to a good life (5.1–5.6 are outcome targets, while 5.a-5.c are targets describing means of implementation).

Target 5.1, *end discrimination against women and girls*, is measured by national statistics: on education, how many women have received different levels of education; on employment of women, in which jobs women are employed less and in which more. Target 5.2 addresses the elimination *of violence against women and girls in the public and private spheres*; the criteria for the measurement of progress are the *proportion of ever-partnered women and girls subjected to physical, sexual or psychological violence by an intimate partner or another person in the previous 12 months*. Target 5.4 seeks to *recognize unpaid care and domestic work*, and measures it through the proportion of time spent by women doing such work. Target 5.5, *ensure women's participation and equal opportunities for leadership in political, economic and public life*, is evaluated through national statistics of women in business, entrepreneurs, salary differences between men and women, access to bank credit, and last but not least, women's representation in political institutions and state government. Target 5.6, *ensure universal access to sexual and reproductive health and reproductive rights*, is measured in terms of the proportion of women who make their own informed decisions regarding sexual relations and contraceptive use and reproductive health care, and the degree to which laws and regulations guarantee full and equal access to sexual and reproductive health care, information, and education. Target 5.a, *give women equal rights to economic resources, access to ownership and control over land, financial services*, is evaluated through the proportion of agricultural population with ownership or secure rights over land, by gender; and share of women among owners or rights-bearers of agricultural land, by type of tenure; consideration is given to the proportion of countries where the legal framework (including customary law) guarantees women's equal rights to land ownership and/or control. Target 5.b, *use of information and communications technology, to promote the empowerment of women*, is qualified via the proportion of women who own a mobile telephone. Finally, Target 5.c, *adopt sound policies and legislation for the promotion of gender equality and the empowerment of all women*, is represented through the proportion of countries with systems to track and make public allocations for gender equality and women's empowerment.[1]

[1] https://sdgs.un.org/goals/goal5#targets_and_indicators

What is immediately identifiable in the targets is the image of an urbanized woman who lives in a time and space dominated by modern facilities and infrastructures. This woman is necessarily a subject of a strong state that can impose the rule of law over its citizens. SDG5 is written after and for the model of Western society. One can argue that Target 5.a also addresses rural communities where land is the main economic asset and means of production. However, the envisioned fair distribution of land requires administrative reforms that are often visualized as government interventions, as is also stipulated in Target 5.c, where sound policies and legislation are promoted as channels for gender equality. Urbanization is said to be a process that accelerates, and policies, including environmental sustainability, need to be preemptive to prepare society to adapt to its negative impacts. SDG5, however, seems to promote the standard of the urbanized modern woman and thus merges preemption with the creation of this same phenomenon, contributing to the very problems that SDGs are supposed to solve: decrease humanity's environmental footprint.

In harmony with the promoted image of modern Western urban woman is the blindness that SDG5 has for community, institutions, and cultural patterns. Following earlier developmental frames, its locus of agency and subject of fairer rules are individual women. As critical research reveals, interventions on the level of individuals can have positive consequences where eventually "ripple effects ... affect communities and institutions. However, individual capacity improvements do not necessarily lead directly and inevitably to increases in gender equality" (Mathews & Nunn, 2019, p. 33). We can observe, thus, that SDG5 is informed by specific ideas of personhood and society that fit best Western neoliberal models of state building, societal and economic progress, grounded in individualism, autonomy, private property, and atomized communities (e.g., Haller, this volume). In this context, it is not surprising that SDGs do not explicitly mention or address Indigenous people or other groups that have been globally granted cultural group rights, as such rights have proven to face popular opposition by dominant society and state officials in many parts of the world, including sustainable development leaders like Nordic countries.[2] In this vein, SDGs fail to acknowledge Indigenous rights. Indigenous scholars and politicians revolutionized the concept of rights in order to struggle against the ongoing injustices of colonialism that persevere the claimed end of colonial states through societal structures and knowledge practices named also as epistemic violence and coloniality (Mignolo & Walsh, 2018; Spivak, 1988). The global Indigenous movement also raises awareness of cultural and historical diversity in the world and of the universal right to self-determination. SDG5 does not reflect the message of multiculturalism that diverse values and sources of human well-being can exist along mainstream Western values. We do not reject these values, or SDG5's important message and positive impacts

[2] A common argument that such opposition voices is that Indigenous people in welfare states already have the same social security rights as other citizens, so group rights that exceed these create injustice and social privilege (Olsson & Lewis, 2002).

on governments worldwide. However, by failing to acknowledge, affirm, and build on culturally specific practices and sources of individual and collective women empowerment in diverse societies, SDG5 has limited capacity to bring all women justice.

In considering the important aspects of women's equality reflected in SDG5, we notice that it fails to promote globally the implementation of women's well-being on two different levels. First, it fails to open for critique colonial structures of knowledge that create our world's dominant axis of inequality. The solution it offers—transform all societies after Western model and values—instead reproduces coloniality and the preconditions for deepening world divisions. Second, it fails to accept cultural diversity and thus misses the chance to learn directly from and talk to Indigenous and rural communities and reinforce those practices that de facto give women power and status in society.

In this chapter, we initiate a debate on the possibilities to decolonize, democratize, and "customize" SDG5. In the limited space we have, we do not start re-formulating the goal. Instead, we substantiate our criticism by bringing to light cultural practices that give women high status and power in their communities. We discuss two societies, with diverse cultures but sharing similar Soviet historical legacies: Kyrgyzstan, and Nivkh Indigenous people on Sakhalin Island, Russia. After briefly introducing the SDG frameworks in Russia and Kyrgyzstan, we sketch some ideas of women's emancipation that characterized the Soviet modernization project and its present legacies that predate and inform SDG implementation in post-Soviet space. Finally, we turn to Nivkh and Kyrgyz communities to show how women negotiate and enact their agency in a fast-changing world. Indigenous women's perspectives on their roles and experiences of the transformations that SDG5 attempts to navigate are a primary focus.

SDG5 and the Conservative Gender Turn in Kyrgyzstan and Russia

Indigenous people of the Russian North and from Kyrgyzstan were both part of the Soviet social engineering experiment. Soviet ideologists used the kinship terms "elder brother" and "younger brother," to distinguish between ethnic Russians and other groups from Siberia or Central Asia, which reflects a hierarchical relationship typical for a colonial system. Post-Soviet Russian Federation and Central Asian countries remain closely connected economically and culturally through Russian language. Both countries demonstrate openness to SDGs, and they have each developed differing institutions for measuring their implementation. The countries' ideologies and legal activities have been discordant, but both have been affected by conservative gender politics, that is, an attempt to popularize "traditional values" that emphasize women's roles in the domestic sphere (Sätre et al., 2023). In line with such policies, the Russian authorities decriminalized domestic violence and disempowered most NGOs working with women's rights (Johnson et al., 2021). War-time state propaganda has taken conservative values to the extreme, promoting prohibitions on abortion, as well as early pregnancies and births.

In contrast, Kyrgyzstan has a continuous commitment to SDGs, including goal 5. We see the country's dedication through certain strategic documents and collaborative efforts with international organizations. Kyrgyzstan's National Sustainable Development Strategy for 2018–2040, although not specifically focused on gender aspects, emphasizes the importance of providing equal opportunities for all citizens, regardless of gender, to realize their full potential.[3] In 2020, Kyrgyzstan presented its first Voluntary National Review (VNR) of the SDGs at the High-Level Political Forum (HLPF) on Sustainable Development. Regarding gender equality, the VNR noted that the Kyrgyz Republic ratified the United Nations Convention on the Elimination of All Forms of Discrimination against Women in 1996 and adopted the Beijing Declaration and Platform for Action. One of the main priorities of the National Development Strategy for 2018–2040 is to provide citizens with equal opportunities to realize their abilities, regardless of their wealth, social status, place of residence, ethnicity, gender, and other differences (VNR, 2020). The United Nations Development Programme (UNDP) in Kyrgyzstan has developed a Gender Equality Strategy for 2018–2022. This strategy aims to strengthen both central and local public administration capacities to implement and monitor existing policies of gender equality. UN Women has published a report titled "Sustainable Development Goals and Gender in Kyrgyzstan," which provides an analysis of the current state and progress in the field of sustainable development and gender equality. This publication utilizes official and departmental statistics, as well as results from the Multiple Indicator Cluster Surveys (MICS) conducted in 2014 and 2018, to inform policy decisions related to SDG implementation. The National Statistical Committee has compiled a compendium titled "Monitoring of the Sustainable Development Goal Indicators in the Kyrgyz Republic," which provides an overview of data on the SDGs. The compendium includes data that can be disaggregated by gender, which is essential for assessing progress on gender equality, but it does not offer an in depth, focused analysis of SDG5 (Monitoring, 2020). These efforts of the government and international organizations perhaps popularize ideas of gender equality, but they have uneven reach to different parts of the country and are often limited to urban centers. Rural populations have different concepts of gender equality and practices for women's empowerment, one of which is the traditional custom of singing *koshok*.

Kyrgyzstan is also experiencing a "conservative turn," which is emphasized in President Sadyr Japarov's first decree, signed on January 28, 2021, "On Spiritual and Moral Development and Physical Education of the Individual." This decree highlights the promotion of "high moral standards," traditions,

[3] https://sdgs.un.org/system/files/webform/sdg_national_commitments_form/21799/national_development_strategy_of_kr_2018-2040_final_eng.pdf

traditional family and values, and a healthy lifestyle (Japarov, 2021). However, the country's religious turn that gained momentum in recent years has even more significant impact on social norms, governance, and public life. A religious radicalization, where many Indigenous practices—especially those traditionally involving women—are becoming forbidden or marginalized. *Koshok*, which was traditionally created and performed by women, is now deemed inappropriate.

Developmental Framework of Women's Equality and Soviet History

Until recently, the developmental epistemological frame has been dominant in national state policies for Indigenous peoples and rural populations. It evaluates women's conditions against modern ideas of social and economic progress. Improved material conditions, access to education and medical care, urbanization, salaried jobs, and public culture should bring women well-being globally. The Soviet Union is an emblematic example both in the elaboration of this epistemology and its implementation.

Soviet policies, first, that of "social enlightenment" since the 1920s, and later of forced collectivization, and "cultural construction" since the 1930s, were justified through a representation of many ethnic people in the Arctic and Central Asia as occupying lower evolutionary stages according to a linear hierarchy of cultural development (Grant, 1995). Nivkh, a Numerically Small People according to Russian classification, numbers 3842 people (Russian Census, 2021 retrieved from https://rosstat.gov.ru/vpn/2020/Tom1_Chislennost_i_razmeshchenie_naseleniya link is not working in Sweden as of 24 February 2026.). Emblematic in the case of Nivkh people is the debate in mid-twentieth century Soviet ethnography at what stage of social development Nivkh society was: according to Shternberg, in late nineteenth and early twentieth centuries, Nivkh people practiced a form of collective marriage that he classified as one of the oldest (Shternberg, 1904). Such ethnographic narratives were instrumental in condemning polygamy and other kinship institutions that have been common among many peoples worldwide. Levirate, sororate, bride kidnapping and bride wealth, and polygamy are just some of these institutions that Soviet authorities criminalized legally in Central Asia. Toward Indigenous peoples of the North, like Nivkhs, who were perceived to occupy a lower stage of social development than Central Asian nomads, the authorities undertook "softer" policies only condemning them and the people who exercise them as "dark" and "primitive", and impediment to modern health and civilization (Vladimirova & Liarskaia, 2023).

In 1930, the dominant ideology of women emancipation proclaimed egalitarianism through equal involvement of women in labor and education and all Soviet women were expected to be members of the party and women's organizations. Soviet emancipation was justified by ethnographers' claims that Indigenous and Central Asian women were oppressed and victims of supposed pre-modern stages of social organization. Many traditional kinship practices were condemned and people who exercised them were stigmatized. The Soviet government accused hunter-gatherer and pastoralist economies of inefficiency, and of depleting the resources on which they depended. The Soviet model of gender equality was based in equal rights to education, property, marriage, and custodianship over children, as well as obligatory employment for both sexes. A variety of social and psychological problems that persevere today among post-Soviet communities originated in this aggressively enforced reformation.

Despite contemporary interest among many Indigenous people in recording and reviving earlier cultural forms, modern views of the family and women's position in society are the norm. Preservation of tradition focuses more on handicrafts, foods, and folklore as sources of national pride. Traditional culture related to old marriage and kinship forms is erased through silence within communities. Russian ethnography, on the other hand, continues to reproduce the narrative of "traditional" Nivkh female roles and models of behavior subjected to male domination using the ethnographic present and refusing coevalness of Nivkh culture (Fabian, 2014; Fadeeva, 2013). Indigenous scholars are being coerced to reproduce the same narrative and none has so far problematized this form of knowledge in the ethnographic literature (Taksami, 1976).

In the context of Soviet ideology, Kyrgyz women were represented in research as subjects of transformation, emphasizing the progressive changes brought by Soviet policies. Ethnography and history of the Soviet period, presented in Kyrgyz Soviet Republic by Russian scholars and their students, including local researchers—both Kyrgyz men and women—offer a roughly uniform description of the role and status of Kyrgyz women before the 1917 revolution: a disenfranchised figure in a patriarchal society, the Kyrgyz woman gained access to education, employment, healthcare, and other social rights and opportunities (Abramzon, 1971).

This transformation is vividly portrayed in Chingiz Aitmatov's famous novella *The First Teacher*, in which the main character, Altynai, is forcibly married off at a young age to a much older man whom she does not know (Aitmatov, 2018). The story highlights the key indicators of a woman's status in the pre-Soviet period: the lack of choice in determining her own fate, early marriage, the bride wealth (*kalyn*), polygamy, and full subjugation to a male relative (father or husband). If not for Soviet power, embodied by the teacher Duishen, Altynai would have remained a powerless and illiterate younger wife. However, thanks to Soviet policies and the courage of Duishen, the first teacher in the village, the perpetrator is punished, and Altynai receives an education, eventually becoming a renowned scholar.

The fate of female characters reflects the general dynamics of change: from powerlessness and women's restricted roles exclusively to the family sphere (as wives and mothers) to modern-day education and opportunities to step into the larger public world. This transformation indeed brought tremendous changes to the lives of Kyrgyz women, who gained the opportunity to participate in political life. The emergence of women as teachers, doctors, and other professionals was often highlighted as an important indicator of these changes.

The images of labor pioneers, teachers, doctors, and cultural workers served as symbols of emancipation and new opportunities (Tatybekova, 1963).

In pre-Soviet Kyrgyz folklore, the role of women is strictly limited to the family, with the functions of mother and wife, and sometimes daughter. However, it is precisely within these roles, complementing and serving the man, that a woman gains social recognition. Such is the case with Kanykey, the fourth wife of the epic hero Manas. An attempt to transcend the family circle, to step beyond the role of a woman defined by her relationship to a man, is made in the epic *Janyl Myrza* and ends tragically. A woman standing independently, not as a mother or wife, is rejected by the Kyrgyz tribes (Aitpaeva, 2014). At the same time, through various studies, we can understand that during the pre-Soviet era, women played a key role in the creation of the range of folklore genres, in the production of certain crafts such as weaving, felt production, and embroidery, which were considered part of strictly female labor that provided for the household's needs (Antipina, 1962). Despite official criticism of religious rituals, many ethnographers noted the role of women in preserving and transmitting traditional folk beliefs, such as the worship of ancestral spirits (Baialieva, 1981).

Ethnographers working within the Soviet paradigm focused on the negative aspects of many traditional practices, often shifting their meanings. For example, in relation to the phenomenon of *kalyn*, they viewed it as a remnant of feudal society. *Kalyn*, or bride wealth, was interpreted as a practice that dehumanized women, turning them into commodities. Later Kyrgyz researchers began to view it as a form of compensation for the labor and expenses of the bride's family, an expression of gratitude for her. During the Soviet period, *koshok* underwent notable adaptations. Unlike certain folk traditions that were strictly prohibited, folklore, including *koshok*, was reinterpreted within the framework of socialist realism. Although not entirely banned, *koshok* experienced significant changes to align with the ideological goals of the era. While traditional *koshok* focused on personal and communal sorrow, "Soviet *koshok*" incorporated themes of socialist realism, focusing the details and realities connected to the regime. This shift reflected the broader trend of integrating folk traditions into state propaganda (Erkebaev, 1999).

During the late 1930s, a significant portion of Soviet folklore consisted of funeral laments for Soviet officials and celebrities. These laments, known as "Soviet laments," were crafted to honor the contributions of the deceased to the state, blending traditional mourning practices with political content (Panchenko, 2005). This transformation of laments from expressions of personal grief to instruments of state ideology occurred across the Soviet Union and exemplifies the broader effort to reshape folk traditions along social structures and economy. These changes aligned traditional practices with the promotion of Soviet values. However, these changes were not universally embraced, and traditional *koshok* reflecting personal suffering and loss continued to exist in certain regions, especially in rural areas (Akmataliev & Egemberdieva, 2016).

Indigenous Framing of Women Empowerment

Indigenous feminist scholars problematize evaluations of Indigenous women's conditions. While they do not entirely dismiss the developmental frame, they question its assertions and universal definition of gender equality. This criticism is grounded in histories of forceful imposition of modernization and dominant society values on communities without their consent. Indigenous thinkers blame crises in families and communities, such as gender violence and other atrocities in Indigenous societies, on the structures and legacies of settler colonialism that have never been effectively questioned by state governments (Behrendt, 1993; Flowers, 2015). They reveal development's hidden agenda as extending state control over Indigenous peoples and promoting and justifying the exploitation of their territories and resources. Developmental goals, including the denunciation of traditional kinship relations and women's roles in society as repressive, are thus coterminous with colonialism. In order to confront it, Indigenous scholars suggest grounding the assessment of Indigenous women's well-being in terms similar to the definition of Indigenous rights: self-recognition, definition, and ownership of Indigenous culture and its transformation (ibid.).

In the Russian Arctic, such ideas are implicit in recent studies that praise women's resistance, involvement in Indigenous politics, and cultural revival (Vladimirova & Habeck, 2018), but often they do not part with dominant ethnographic accounts of Indigenous cultures. Indigenous empowerment is described predominantly with the terminology set by nation states—in legal provisions and number of Indigenous representatives in different bodies of state governance. In Kyrgyzstan, there is a tension between "liberal" and "traditional" ideologies, and women actively contribute to this dynamic. Kyrgyz female researchers, both within the country and abroad, often align with mainstream feminist ideas in their writings and publications (Ibraeva et al., 2024). In contrast, Kyrgyz female traditional activists rarely publish their ideas, but in informal discussions notice that feminism undermines "traditional values" and destabilizes Kyrgyz society. Considering that *koshok* has traditionally been and continues to be performed by women, it can be viewed as a lens for indigenous female thought and, to some extent, as a form of indigenous feminist storytelling.

Women's Perspectives on Empowerment in Nivkh and Kyrgyz Communities

Soviet policy was grounded in educating Indigenous elites that aspired to live by the values of Soviet modernity and be loyal to the regime. Besides motivating women to acquire higher education, Soviet policy directed Indigenous women into jobs and positions that are typically urban and provide better opportunities for networking with dominant social groups. This displaced and alienated women from some economic activities and spaces, like, for example, reindeer husbandry, and nomadism, but provided them with symbolic capital and high social status. The Nivkh intelligentsia is composed predominantly of women who work in administrative offices, educational and medical institutions. Nivkh teachers are spoken about with great respect. In an interview in Nogliki, a Nivkh teacher Natal'ia Vasil'evna, with 30 years' work experience, elaborated strategies that could help Indigenous language revitalization. Her suggestion was the publication of more textbooks in Nivkh with contemporary texts. As possible authors she named the famous writer Vladimir Sangi, who stands out as the most respected Nivkh intellectual. Next to his name, however, she listed three female teachers of different ages who in her view were all talented enough to produce excellent textbooks. Natal'ia graduated from the State Pedagogical University in Magadan in 1981, but her childhood was spent in several smaller villages where the family was forcefully moved in the 1960s, when numerous small enterprises were closed and their resources and employers moved to bigger villages.

In the 1940s Nivkh women were recruited into fishing enterprises and agriculture to replace the male labor force, which had been redirected to the WWII fronts. Older women remember with pride the time and the praise and public attention they received (Grant, 1995). Soviet and Russian ethnographers speak of traditional Nivkh gendered labor division in terms like male hunters and female householders (Fadeeva, 2013). In their narratives, women's economic roles were subjugated and created life-long dependency on male figures such as father, husband, and brothers. Ethnographers assert the authority of this view by mentioning that Nivkh women's positive evaluation of their traditional subjugated positions is unreliable, since their narratives are biased by their culture's narratives and values (ibid.).

Nivkh fishing economies were collectivized in the 1930s and people were forced to salaried jobs. By the 1970s, industrial fishing replaced the cooperative system and Nivkh people were excluded from fishing altogether, their traditional subsistence economy. Today, Indigenous legislation in Russia provides a small fishing quota for subsistence needs for each individual (100 kg salmon per year). However, most coastal areas were not open for Indigenous fishing in 2009. One coastal area that was still open was the Nyvo Bay on southern side of the island, where several families were gathered in the autumn to catch their quotas of fish. Fish could not be officially sold according to regulation, but through unofficial deals many families earned in this way their annual income. Both men and women were involved and complemented each other. The work women on the bay conducted was described in positive terms and with pride. Cleaning fish, preparing them for marketing, or drying them for winter were shown to anthropologists as important skills equal in value to the making and setting of fish traps, and sea mammal hunting, conducted by men. Nivkh women described their fishing work and the income from it as significant for their roles in family and society. The life in the camp, in the words of one woman, is particularly good for children and youth, as they come to know how to procure food and live in the natural environment (Fig. 12.1).

Fig. 12.1 Fishing on the Nyvo Bay. (September 2009. Photo Vladislava Vladimirova)

A narrative that many female respondents from different generations reiterated in different contexts, both intellectuals in the village and fishers in the camp, is the role of women in countering the alcohol drinking of men. As Liudmila, an elderly woman commented, it was women who insured that families survived and life continued after the mid-twentieth century, when Nivkh men lost their role and jobs in fishing and many could not handle this cultural loss. Women take pride in their power to resist and bring up families even in the toughest of circumstances.

Indigenous women with higher education and urban and office-based jobs have been significant force behind the Indigenous ethno-political revival movement in Russia since the 1990s. Larger numbers of women are engaged in regional and local organizations, while leadership of the main organizations is male-dominated. While careful analysis of the implications that such gender division might have is important, the active participation of women in ethno-political and cultural work and events provides a field for cultural communication, knowledge exchange, shared community experiences, creativity, and representation of women that needs research attention. Within the realms of family histories, narration and folklore, handicraft, and ethnic cuisine, women have great liberty to narrate, revive, and reinterpret Indigenous culture in the present. These realms are perhaps least governed by the contemporary extractive political regime in the Russian Federation that severely limits Indigenous rights to natural resources. Mastery and active roles in the realms listed above are thus sources of respect and authority in Nivkh communities. The majority of women interviewed speak with respect and warmth about *masterícy* (in Russian), skilled handicraft practitioners, and folklore performers. Village activists who are involved in the everyday work of procuring resources and organizing such activities for the community are influential at a higher level than Nivkh women representatives at the regional administration in Sakhalin, for example. According to SDG5 Target 5.5, women's equal participation in institutions of governance is an indicator of women leadership and empowerment in society. According to Nivkh, however, such representatives, female as well as male, are to a high degree restricted by power structures. They have little leverage, resources, or desire to promote the interests of Indigenous people and create positive changes. In the words of Katya, a respondent from the village of Nogliki:

> Law is a wall, it conceals. I do not even try to go behind the wall. We do not have Indigenous leaders that can go behind this wall.... if one can go in the interest of our people, not just one's own, people will follow. Now we are all passive, doing things only for ourselves, or nothing. Those who are at the highest positions, none takes their words seriously. Our Queen did not for a long time even have a right to vote. She could attend meetings but had no right to speak.

People called the Nivkh Queen the representative of Indigenous people at the regional government in Yuzhno-Sakhalinsk who at the time was a woman. She, like Indigenous female representatives with high-status positions in other regions of the Russian Arctic, has been the subject of criticism within the community. One common contestation concerns how representatives are being elected within Indigenous organizations and if they really represent their people. On many occasions, Indigenous individuals question whether such representation is founded on democratic principles and attribute the selection of representatives to political and economic processes that are not transparent to the majority and often are influenced by favoritism from municipal and regional government figures. High-status political positions for women do not necessarily allow them to become Indigenous community leaders nor increase Indigenous women's empowerment.

Field research conducted in October 2024 in one of the northern regions of the Kyrgyzstan, Talas, revealed that *koshok* was banned by clerics and "strict" followers of Islam, silent opposition between proponents of a new wave of Islamism and traditionalists, and active debates where both sides defend their views on social media. Talas is known as a region that preserves traditions, and we visited a community organization called "Talas ayimdary" that brings together creative women engaged in singing, embroidery, and felt crafts. Among them, we found women who still uphold the tradition of *koshok*. We met with nine of these women, whose ages range from 70 to 89 years. All of them inherited the *koshok* tradition from their mothers or grandmothers. They combine the recitation of *koshok* with other practices, such as writing poetry in a folkloric style, creating felt products, and traditional storytelling. One of them is Kalemkash Alikeeva, the mother of the renowned akyn Malik Alikeev, who shared her sorrowful experience. Malik passed away in 2023, and at the age of 92, Kalemkash had to mourn the loss of her second son. Kalemkash explained that *koshok* emerges from deep grief. "When your heart is filled with sadness, *koshok* comes," she said. A mother goes to her son's grave, addressing him through a lament—a song overflowing with tears. Through this mournful expression, she finds a small measure of solace for her overwhelming sorrow. What is significant about Kalemkash Apa's case is that her ability to recite *koshok* and her talent for poetic improvisation keep her actively engaged in her community, even at her advanced age. She is regularly invited to funerals, memorials, and other important events where her skills are highly valued. This demonstrates how the demand for traditional knowledge can support and empower an individual woman.

Farida Chynabayeva also highlights the importance of being needed and valued within the community. Her reflections, along with her account of other women who practice *koshok* (*koshokchu*, in Kyrgyz), shared on Aigine's Facebook page, sparked a wave of support for the tradition. Women shared their experiences of how and why they embraced *koshok*, emphasizing its emotional and spiritual significance. Few mentioned its economic aspects, yet they do exist. Funerals, traditionally, come with expenses—such as the cost of sacrificial animals, food for attendees, rituals, and *koshok*. However, there is now a noticeable shift toward channeling these expenses primarily into religious practices. For instance, the deceased is taken to the mosque for *janaza aytuu* (Islamic funeral prayers), and related expenses also flow there. This shift not only diminishes the potential, albeit modest, economic role of *koshokchu*, but it also marginalizes women from a crucial life ritual. Women are central to the life cycle—they welcome a person at birth and, through their lament, bid them farewell to the next world. There is now an effort to exclude women from this vital cycle of life and death.

Koshok is being deliberately suppressed by strong current efforts to remove it from public discourse. During the Soviet era, it was suppressed due to atheistic ideologies, and today it is sidelined for religious reasons. However, given the profound psychological qualities of *koshok* and its connection to what is now referred to as emotional intelligence, it is extremely difficult to erase it entirely. *Koshok* is not merely a lament—it is a deeply significant ritual that helps the grieving express their sorrow and let go of the departed. Through the recitation of *koshok*, the life, achievements, and journey of the deceased are honored, preserving their memory in the community while providing a way to process and accept the loss. In this way, *koshok* becomes an essential tool for emotional and spiritual healing for the performer, for the family and relatives of the deceased, and for the whole village. This tradition is primarily maintained by the older generation, for whom *koshok* remains a living memory and cultural anchor. Women play a central role in its performance, not only preserving an important tradition, but also asserting their cultural significance and emotional strength. By performing *koshok*, women take on an active role in the communal process of mourning and healing, further solidifying their position as emotional anchors within their communities. This empowering aspect of *koshok* has made it a target for suppression. The practice gives women a voice, a platform, and a vital role in significant life rituals—elements that some view as challenging to the shifting dynamics of religious and social authority. As efforts to suppress *koshok* grow, they risk erasing not only a rich cultural heritage but also a critical avenue of empowerment for women. *Koshok*, like many other traditional practices, was officially unpaid but publicly recognized and contributed to the well-being of Kyrgyz women. By serving as a culturally significant tool for emotional expression and social support, *koshok* recognized a vital role in the communal life of Kyrgyz women.

Conclusion

Analysis grounded in women's evaluations of their situation raises questions about both Soviet assessment of emancipation through modernization and recent developmental frames like SDG5. Women's power and well-being do not evolve universally from high-status administrative positions, economic affluence, and urban facilities or use of technology that SDG5 attempts to capture through numerical measurement. Women in different communities and cultures suffer from specific forms of oppression and, respectively, develop diverse strategies to negotiate their agency, to get higher status and power, and most importantly, to live good and meaningful lives. Their efforts have to navigate dynamic social institutions, religious authority, national and global structures of governance and knowledge. At the same time, they can use creatively opportunities and resources provided by national and international frames, cultural norms, and traditional practices. As our ethnography shows, societal positions premised on education, job, high income, and urban location do not necessarily translate into women power and equality. Women hold other important roles in communities too, as handicraft, householding skills, and ownership over folklore forms like *koshok*.

SDG5 can provide knowledge and legal and infrastructural support in some instances, but its metrics, visions, and instruments are constrained by the limits of the perspectives, positionality, and live experiences that it presents and deems universal. While its ambition is to democratize and expand notions and practices that increase justice and prosperity while decreasing humanity's environmental footprint, it replicates some mistakes and blindness that models like the Soviet modernization has shown: reproduction of colonial frames of knowledge and discrimination that reject diversity and freedom, and split the world into categories of quality. Ranking of states' implementation of SDG5[4] reinforces stereotypes and is equally counterproductive for sustainable development "leaders" like, for example, Sweden, as for laggers behind, as it hinders recognition of continuing inequalities by concealing them from critical scrutiny. It prevents even an opportunity for public debate and political action and it silences affected victims and even deems them non-existent. Associating on the other hand "bad practices" in relation to women with particular nations,

[4] https://dashboards.sdgindex.org/rankings

societies, and ethnic groups is a strategy to reject their universal character and blame particular groups or cultural elements for them. Attributing atrocities to a group, a knowledge system, or a culture is a way to avoid responsibility for global economic and environmental injustices and political contempt and dominance, as well as colonial frames of knowledge, that are the usual context for women inequality and suffering.

Acknowledgments The cooperation that led to the co-authored chapter was facilitated by CARSI—Project Number: 101086415, Caucasus and Central Asia Research Social Innovation: Development Assistance, Innovation and Societal Transformation. Fieldwork among Nivkh people was part of the project CLUE Dynamics of Circumpolar Land Use and Ethnicity, funded by the American National Science Foundation (grant number 213665). We thank Nathan Light, Elena Liarskaia, and Maria Liarskaia for their comments and help with the chapter.

References

Abramzon, T. M. (1971). *Kirgizy i ih etnogenetičeskie i istoriko-kul'turnye svjazi [The Kyrgyz and their ethnogenetic and historico-cultural ties]*. Russian Academy of Sciences.

Aitmatov, C. (2018). *The first teacher*. Book baby.

Aitpaeva, G. (2014). Kod Manasa kak odin iz glavnykh kodov perestroiki [Code of Manas as a main code of perestroika]. *Newspaper of the Kyrgyz-Russian Slavic University, 14*(3), 7–11.

Akmataliev, A., & Egemberdieva, S. E. (2016). *Kyrgyz el Koshoktoru [Kyrgyz Folk Lamentations]*. Print-ėkspress.

Antipina, K. (1962). *Features of material culture and applied art of southern Kyrgyz*. Academy of Sciences of Kyrgyzstan.

Baialieva, T. (1981). *Religioznye perezhitki u kirgizov i ikh preodolenie [religious survivals among Kyrgyz and overcoming them]*. Ilim.

Behrendt, L. (1993). Aboriginal women and the white lies of the feminist movement: Implications for aboriginal women in rights discourse. *Australian Feminist Law Journal, 1*(1), 27–44.

Erkebaev, A. (1999). *Maloizuchennye stranicy istorii kirgizskoi literatury*. Zheka Limited.

Fabian, J. (2014). *Time and the other. How anthropology makes its object*. Columbia University Press.

Fadeeva, E. (2013). Status i rol' zhenshchiny v obshchestve i sem'e korennykh narodov Nizhnego Amura: Tradicii i sovremennost [the status and role of women in indigenous family and Society in Downriver Amur: In tradition and at the present]. Zhenshchina V Rossiiskom Obshchestve, 4, 65–76.

Flowers, R. (2015). Refusal to forgive: Indigenous women's love and rage. *Decolonization: Indigeneity, Education & Society, 4*(2), 32–49.

Grant, B. (1995). *In the soviet house of culture: A century of Perestroikas*. Princeton University Press.

Ibraeva, G., Ablezova, M., Danshina, A., & Elias, M. (2024). Developing gender transformative approaches to strengthen women's rights in Kyrgyzstan. Foundational gender analysis. *Global Initiative for Gender Transformative Approaches for Securing*

Women's Resource Rights (WRR), 7. Retrieved from https://cgspace.cgiar.org/items/8ae7ca79-783d-4fb7-8321-e7a533696a95

Japarov. (2021). Decree UP1 on the spiritual and moral development and physical education of the individual, UP1, President of the Kyrgyz Republic (2021).

Johnson, J. E., Novitskaya, A., Sperling, V., & Sundstrom, L. M. (2021). Mixed signals: what Putin says about gender equality. *Post-Soviet Affairs, 37*(6), 507–525.

Mathews, H., & Nunn, M. (2019). Women on the move: Can we achieve gender equality by 2030? Chapter two. Leave no one behind: Time for specifics on the sustainable development goals, 1–18.

Mignolo, W. D., & Walsh, C. E. (2018). *On decoloniality: Concepts, analytics, praxis.* Duke University Press.

Monitoring. (2020). *Monitoring of the Sustainable Development Goal Indicators in the Kyrgyz Republic 2014–2018. A statistical compendium.* The Statistical Committee of the Kyrgyz Republic.

Olsson, S., & Lewis, D. (2002). Welfare rules and indigenous rights: The Sami people and the Nordic welfare states. In J. Dixon, J. Dixo, & R. Scheurell (Eds.), *social welfare with indigenous peoples* (pp. 155–199).

Panchenko, A. (2005). Kul't Lenina i "sovetskii" fol'klor [Lenin's cult and "soviet" folklore. In Russ.]. In *Odissei. Chelovek v istorii. Vremya i prostranstvo prazdnika [odyssey: Man in history: Time and space of the celebration]. Moscow* (pp. 334–366).

Sätre, A.-M., Vladimirova, V., & Gradskova, Y. (2023). Post-soviet women: New challenges and ways to empowerment-introduction. In A.-M. Sätre, Y. Gradskova, & V. Vladimirova (Eds.), *Post-soviet women: New challenges and ways to empowerment* (pp. 1–26). Springer International Publishing.

Shternberg, L. (1904). Giliaki. Etnograficheskoe *Obozrenie* (1), 1–42.

Spivak, G. C. (1988). Can the subaltern speak? In N. Carry & L. Grossberg (Eds.), *Marxism and the interpretation of culture* (pp. 271–313). University of Illinois Press.

Taksami, C. (1976). Predstavleniia o prirode i cheloveke u nivkhov [Perceptions of nature and humans among Nivkh people]. In *Humans and nature in the religions of indigenous people of the north and Siberia (second half of 19c–the beginning of 20c)* (pp. 203–216).

Tatybekova, Z. (1963). *Raskreposchenie zhenshchiny-kirgizki Velikoi Oktiabr''skoi socialisticheskoi revoliuciei (1917–1936) [Emancipation of Kyrgyz women by the great October socialist revolution (1917–1936)].* Frunze.

Vladimirova, V., & Habeck, J. O. (2018). Introduction: Feminist approaches and the study of gender in Arctic social sciences. *Polar Geography, 41*(3), 145–163.

Vladimirova, V., & Liarskaia, E. (2023). 'They Beat Us, We Fly': Indigenous activism among women in the Russian north. In A.-M. Sätre, Y. Gradskova, & V. Vladimirova (Eds.), *Post-soviet women: New challenges and ways to empowerment* (pp. 247–269). Springer International Publishing.

VNR. (2020). *Voluntary National Review on the Implementation of the Sustainable Development Goals in the Kyrgyz Republic.* Retrieved from Bishkek.

CHAPTER 13

More-Than-Human Impacts of Wind Power Development in Finland

Anna Niia Varfolomeeva

Introduction

The 2023 Helsinki Biennial program included a video installation by Sámi visual artist Matti Aikio[1] that intertwined footage of migrating reindeer with wind turbines gradually emerging from the landscape. At times, the bodies of the animals and the turbines nearly merged, blurring the boundaries of both until the human structures faded away. While these human structures and the more-than-human landscapes temporarily coexist in the installation, their potential reconciliation on the land remains an open question for many Sámi and northern residents (Retter, 2021; Figari et al., 2024).

Embarking on this question, this chapter discusses the more-than-human impact of achieving Sustainable Development Goal (SDG) 7 "Affordable and clean energy" in Finland, focusing on wind energy development. It argues that to advance the implementation of SDGs in an equitable and just way, it is vital to attend to the influences of wind energy development on more-than-human landscapes, discussing both its destructive and creative potentials. I view landscapes here as emerging through human and more-than-human relations and histories, shared activities, and practices (Tsing, 2005, Tsing & Lassila, 2017).

[1] Matti Aikio: https://helsinkibiennaali.fi/taiteilija/matti-aikio/

A. N. Varfolomeeva (✉)
Biodiverse Anthropocenes Program; Faculty of Education and Psychology, University of Oulu, Oulu, Finland
e-mail: Anna.Varfolomeeva@oulu.fi

R. Toivanen et al. (eds.), *Decolonizing the Sustainable Development Goals*,
https://doi.org/10.1007/978-3-032-14263-4_13

Consequently, when speaking about communities, I see them as encompassing human and non-human actors, including animals, plants, and industrial objects.

Earlier studies have demonstrated that the effects of climate change are especially noticeable in the northern regions and among socially vulnerable populations (Vincent, 2020). Furthermore, the Russian invasion of Ukraine in February 2022 highlighted the need to develop new energy infrastructures (Berling & Surwillo, 2022). In this challenging situation, the UN Sustainable Development Goals (SDGs) and the European Green Deal are seen as vital instruments for climate change mitigation and reducing inequality (Wolf et al., 2021). Finland—as one of the frontrunners in achieving SDGs—prioritizes strengthening democracy as an important factor in implementing the 2030 Agenda for Sustainable Development (United Nations, 2015), an action plan for the promotion of sustainable development in 2016–2030. In his opening words at the Rovaniemi Arctic Spirit Conference 2017, Timo Soini, Finland's Minister of Foreign Affairs at the time, stated: "The implementation of Agenda 2030 focuses on carbon-neutral and resource-wise Finland and non-discriminating, competent and equal Finland. These two areas are strongly interdependent – they cannot be achieved without each other."[2] There are nevertheless tensions between Finland's aims to combat climate change and protect democratic institutions; these tensions are strongly pronounced in the field of renewable energy development.

The rapid growth of wind farms and mineral extraction for electric vehicle batteries in Sápmi and other northern regions violate residents' land rights, negatively impacting reindeer grazing and potentially damaging Sámi holy places (Zilliacus, 2022). Because of their extensive land use, renewable energy projects threaten the Sámi traditional way of life, particularly for Sámi youth who will face the long-term impacts of the green transition (Seipiharju & Oleander-Turja, 2023). Such tensions could deepen divisions in Finnish society and further exclude local stakeholders from decision-making (Toivanen & Cambou, 2021). Building a strong partnership relationship between corporations and local communities, especially regarding Indigenous rights and environmental protection, is critical to the success of extractive industries (Poelzer et al., 2021). To support the growth of sustainable clean energy industries in Finland, it is thus crucial to thoroughly study local concerns and perspectives.

The chapter aims to contribute to the research on local perceptions of wind infrastructure, seeing the inequalities produced by renewable energy as a more-than-human issue. It examines how wind energy infrastructure shapes local connections to the land and its more-than-human inhabitants. It argues that the development of wind energy may deepen existing societal inequalities, affecting human and non-human residents. Therefore, there is a possibility that wind energy development will lead to conflicts between SDG 7 "Affordable

[2] Full text of the opening remarks of Minister Soini at the Arctic Spirit Conference, November 15, 2017: https://um.fi/speeches/-/asset_publisher/up7ecZeXFRAS/content/ulkoministeri-soinin-avajaispuhe-arctic-spirit-konferenssissa/35732

and clean energy" and SDG 10 "Reducing inequalities." SDGs have been criticized for their anthropocentric nature and their neglect of the welfare of non-human animals (Kopnina, 2020; Torpman & Röcklinsberg, 2021). In a study conducted in 2018 at the Swedish University of Agricultural Sciences, the participants were asked to rank SDGs in terms of their applicability for animal welfare; SDG 7 received one of the lowest scores, i.e., the participants did not see it as related to the welfare of animals (Keeling et al., 2019). This chapter argues that there are direct implications of SDG achievement for animals and other non-human beings and acknowledges the need for further analysis of the more-than-human impact of SDG 7.

To analyze local responses to wind energy development in Finland, this chapter centers on Finnish-language social media analysis, primarily focusing on Facebook groups devoted to wind energy–related debates. Additionally, it analyzes Finnish local newspaper publications and accompanying illustrations covering the topic of wind energy development. These preliminary results aim to stimulate further discussion on the role of wind energy infrastructure as material and symbolic elements of human-landscape relations, embedded in earlier extractive modes and interactions with more-than-human actors.

Wind Energy Development in a Resourcescape

A substance becomes a natural resource only with the human act of appropriation, which constitutes the resource's symbolic "birth" (Ferry & Limbert, 2008). It is possible to trace the "birth" of wind as a resource back to the ancient world. The first windmill aimed at generating electrical power was constructed in 1888 (Kaldellis & Zafirakis, 2011; Johansen, 2021). As wind energy developed, it was closely embedded in societal and political changes, reflected in varying levels of public support. Before the 1980s, however, wind energy development focused on installing individual wind turbines. Since then, attention shifted to the idea of maximizing energy generation by using an array of wind turbines (Gipe, 2018). During this period, discussions around wind infrastructure were largely intertwined with the anti-nuclear movement and growing environmental awareness. As the number of wind turbines and wind power technologies grew in the 1990s and 2000s, cases of local resistance against wind farm construction also became more frequent (Johansen, 2021).

Although wind power development in Finland began only in 2012—later than in many other European countries—the construction of wind turbine generators has grown rapidly. By the end of 2023, a total of 1601 wind turbine generators had been installed in Finland, according to the Finnish Wind Energy Association (*Suomen Tuulivoimayhdistys*).[3] Two years prior, at the end of 2021, there were around 1000 turbines.[4] The growth of wind infrastructure in Finland is conspicuous. Finland's demand for renewable energy is also

[3] https://tuulivoimayhdistys.fi/en/wind-power-in-finland-2/about-wind-power-in-finland
[4] Finnish Ministry of the Environment: https://ym.fi/en/wind-power-construction

growing. According to the Sustainable Development Report 2024, which tracks the performance of all 193 UN Member States on the 17 Sustainable Development Goals, Finland has achieved SDG 7. The growth of the renewable energy share in the country's energy consumption is also noted in the report.[5] The potential negative impacts of renewable energy growth on northern and Indigenous communities are not mentioned in any of the report's indicators. However, as the Nordregio report 2023 suggests, top-down renewable energy projects often dismiss the needs and wishes of Nordic rural residents who do not act as beneficiaries of green energy but rather as tools in achieving national aims (Salonen & Tomrén, 2023).

In this chapter, I situate wind energy development within a larger context of community formation. Wind turbines affect the landscape's visual aesthetic due to their large size and their influence is more visible in areas that are less affected by human activity. In addition to these visual impacts, wind infrastructure impacts the interactions between community members too. An analysis focusing on the materiality of wind as a resource and its potential to produce new relations could allow us to better understand the complexity of local attitudes toward renewable energy development (Choi, 2022).

The chapter explores the interconnections between different forms of resource extraction, building on the concept of *resourcescape* developed in my earlier work (Varfolomeeva, 2023). Drawing on Arjun Appadurai's notion of the multi-layered "-scape" (1996) and related terms like "energyscape" (Kaisti & Käkönen, 2012; Lempinen, 2018) and "minescape" (Ey & Sherval, 2016), the chapter highlights how extractive industries—often seen as opposing other landscape interactions—are deeply intertwined with activities such as hunting, fishing, and tourism. The resourcescape concept highlights how landscapes are perceived and acted upon as valuable in multiple ways—e.g., their aesthetic, recreational, touristic, or hunting and fishing value. These aspects are often interconnected, and the same landscape may be perceived differently. As Sari Janhunen (2018) shows, local residents tend to view wind energy more favorably than second-home owners. For example, only 10% of residents in Ruokolahti, in eastern Finland, felt the landscape would be "ruined" by wind turbines, compared to a majority of second-home owners (Janhunen et al., 2014). This difference may stem from second-home owners' stronger attachment to the landscape's aesthetic and recreational value, as they seek a "nature-based" experience free from technology.

Although wind energy and other means of renewable energy extraction are often juxtaposed with hydrocarbon and mineral extraction, they nevertheless are part of the extractivist chain and similarly bear the potential of negative environmental and societal consequences (Dunlap & Marin, 2022). Wind extraction often continues local industrial histories (Össbo, 2023) or happens alongside other forms of extraction, resulting in "energy accumulation"

[5] Sustainable Development Report 2024, Finland: https://dashboards.sdgindex.org/profiles/finland/indicators

(Günel, 2022, p. 690) rather than the energy transition. Earlier experiences with other types of resource extraction may inform the local attitudes of acceptance or opposition to wind energy. Thus, the factsheet published by the Sámi Council notes that wind energy development in Indigenous territories contributes to negative effects already produced by forestry, hydropower, and mining (*Renewable Energy*..., 2022). To better understand the complexity of local attitudes toward wind energy development in the North, it is important to analyze wind infrastructure not as standalone features of the landscape, but as interconnected elements embedded in multiple forms of interactions between humans and more-than-human actors, as well as place-based industrial histories.

Methodology

This chapter presents initial findings from a social media analysis of two Finnish-language Facebook groups focused on wind power opposition: *Tuulivoima* (Wind Energy) and *Tuulivoimalapolitiikka* (Wind Energy Politics). *Tuulivoimalapolitiikka* is an open group and posts are public, while *Tuulivoima* is private, requiring membership approval (granted without any additional questions asked). As I was finalizing this chapter in November 2024, *Tuulivoimalapolitiikka* had 1.4 k members and *Tuulivoima* had 951. Both groups are active, with at least one new post daily. Both are also critical of wind energy development in Finland. However, members often reference more positive perspectives on wind energy from other media, allowing for a range of opinions to be tracked within the discussions.

The use of social media in research poses several ethical concerns, such as the blurred line between public and private, the potential for anonymity breaches, as well as the risk of harm, especially when engaging with sensitive data (Townsend & Wallace, 2016). The changing social media policies regarding data privacy, as well as the general lack of information about privacy policies and data access among users, contribute to the problem (Taylor et al., 2023). For the sake of transparency, I wrote posts presenting myself as a researcher working on the topic of wind energy development in Finland when I joined both Facebook groups in the spring of 2023. In order to not reveal the names of any of the group members, this chapter does not quote any specific posts or discussion threads, but instead retells their contents in English (the groups only use Finnish). The social media analysis is still ongoing, and therefore the aim of this chapter is not to present a full overview of the discussions, but rather to highlight several themes that I see as instrumental for investigating the more-than-human impact of achieving SDG 7.

In addition to social media research, I refer to multiple Finnish regional newspaper publications and their accompanying illustrations that are devoted to the impact of wind energy on animals. The chapter quotes from *Kaleva*, the largest local newspaper of the Northern Ostrobothnia region; *Valkeakosken Sanomat* (*Valkeakoski News*), the newspaper of Valkeakoski municipality located in the western part of Finland; and *Maaseudun Tulevaisuus* (*The Rural Future*), published in Helsinki by the Central Union of Agricultural Producers and Forest Owners. While my research does not center on the analysis of newspaper publications, I see them as a valuable additional source that can provide context for the themes raised in social media discussions. When working with published sources, I provide direct citations translated by myself from Finnish to English.

In the following sections, I will focus on two interrelated themes regarding the more-than-human impacts of wind energy. First, I will analyze how wind energy development is portrayed in social media and newspaper articles as a potential threat to animal well-being and the aesthetic value of landscapes. Second, I will examine how wind power activism is often rooted in pre-existing and familiar more-than-human connections, such as attitudes toward invasive species. These themes highlight the complexity of wind power perceptions among affected residents. They also underscore the dual role of extractive industries in human-landscape relations, where emerging technologies may simultaneously disrupt existing more-than-human connections and contribute to the formation of new links or alliances, thus contributing to the formation of a new resourcescape.

Wind Infrastructure as a Threat to Landscapes and Wildlife

The images posted in the Facebook groups under study often contrast with the promotional photos distributed and shared by wind power agencies, such as the Finnish Wind Power Association (*Suomen Tuulivoimayhdistys*) founded in 1988.[6] For example, in spring 2024 the front page of the Association's website featured several promotional images, including two placed side by side. The photo on the right showed a wind turbine from above, surrounded by forest, with little human activity around it, except for a single car on a wintry road. Though the turbine dwarfs the surrounding forest, this is not immediately apparent due to the photo's angle. The photo on the left showed three cranes flying peacefully alongside wind turbines, seemingly unaffected by the turbines' movement or noise.

In contrast, the photos uploaded by the members of *Tuulivoima* and *Tuulivoimalapolitiikka* usually picture wind turbines from below, e.g., captured by a person standing on the ground or passing by in a car. In these images, the wind turbines often appear as very tall and even daunting landscape features in contrast to the nearby trees or houses. Sometimes the group members post photos from the construction sites, showing how even an unfinished wind turbine towers over its surroundings. These photos present the

[6] In August 2024, *Suomen Tuulivoimayhdistys* was renamed *Suomen Uusiutuvat* (The Renewables Finland), as it now focuses on industrial solar power as well as industrial wind power. Source: https://lahienergia.org/suomen-tuulivoimayhdistys-on-nyt-suomen-uusiutuvat/#

resourcescape of wind power as an inhabited space with roads and settlements next to wind infrastructure.

In many cases, group members also post photos representing the interactions between animals and wind infrastructure, although these interactions are presented as far less peaceful than in the Finnish Wind Power Association's photos. For example, after my self-introduction post in the group, a female member of *Tuulivoima*, with whom I had exchanged several messages, sent me photos taken a year ago near the city of Oulu in Finland. The photos pictured a family of white-tailed deer (*Odocoileus virginianus*, in Finnish *valkohäntäpeura*) next to a wind turbine. This species of deer is native to North, Central, and South America and was brought to Finland in the 1930s as a hunting animal.[7] The group member mentioned she had not seen the deer family recently and wondered if their absence was due to the nearby wind infrastructure, suggesting that it could be dangerous for the animals. At the same time, the photo represents a (temporary) coexistence between wind infrastructure and animals, similar to the installation by Matti Aikio, with which I opened this chapter. However, in Aikio's installation, the wind turbines eventually disappear, leaving the reindeer to graze. The close interaction of animals and wind infrastructure is presented as a time-limited condition, implying that while wind infrastructure is a temporary phenomenon, reindeer are timeless. My Facebook interlocutor's worry was the opposite; namely, that the white-tailed deer family disappeared from their grazing grounds leaving only the newly emerging technologies.

A similar issue was also discussed in a post that appeared in the *Tuulivoima* group in 2021. There, a group member posted a photo of an eagle (in Finnish, *kotka*) wounded by a wind turbine, stating that public statements regarding the safety of the wind infrastructure are false and that wind turbines' blades are being "washed with blood." Another user commented on this post, claiming that no eagle would be safe flying next to a wind turbine. This discussion juxtaposes with promotional images such as the one used by *Suomen Tuulivoimayhdistys*, which show birds flying right next to wind turbines as coexistent. As the discussion demonstrates, this coexistence is doubted by the members of social media groups residing in the areas affected by wind power construction.

In some of the posts, wind turbines and animals or birds are presented in direct opposition to each other. Thus, in the comments thread following the post about the wounded eagle, a group member remembered a case when the protection of a golden eagle's nest (*Aquila chrysaetos*; in Finnish, *maakotka*) became a reason to pause wind power construction in a settlement in central Finland. The group member concluded, "If only we had more golden eagles!" The golden eagle is indeed an endangered species in Finland, and even though its numbers have been slowly increasing in recent years, it is still under threat for a variety of reasons including habitat destruction. In 2023, anyone

[7] Source: Finnish Biodiversity Info Facility, https://laji.fi/taxon/MX.47629

reporting a golden eagle's nest in Finland received a reward of 100 euros.[8] This statement, wishing for more golden eagles in Finland, could thus be read on several levels: as a wish for the endangered species to develop further and as an appeal for a more-than-human alliance against new technologies.

Such an alliance is also implied in an article from the *Kaleva* newspaper published on December 10, 2022. The article examines the case of the halted wind energy project in the Kajaani region in the north of Finland. It is metaphorically titled "The wolf is increasingly in the path of the windmill" (in Finnish, "*Susi on yhä useammin tuulimyllyn tiellä.*").[9] The article is accompanied by an illustration presenting a howling wolf surrounded by fir trees, which create a symbolic "shield" protecting the animal from the wind turbines situated around. The article and its accompanying illustration present the relations between animals and human-built structures as strongly oppositional. The wolf's habitat is demarcated and clearly separated from the wind turbines' area. The wolf is in the active pose; its howl is a potential protest against wind power expansion. In another media article covering the story, this one in *Maaseudun Tulevaisuus*, the title reaffirms this opposition: "Wolves Destroyed the Wind Power Project in Kajaani."[10] Like the *Kaleva* article, this article presents the wolf as an agent taking an active position against wind infrastructure construction. The accompanying illustration pictures a wolf biting a broken cable leading toward several wind turbines standing on the horizon. Like the illustration in *Kaleva*, this drawing presents wolves as participants in local activism against wind turbine construction. It implies that wolves do not simply prevent wind power expansion by their existence and status as a protected species, but take specific actions to damage wind energy infrastructure.

Historically, Finnish attitudes toward wolves have been complicated, and the wolf is often viewed as "the most controversial animal in Finland" (Komi & Kröger, 2023, p. 501). While wolves are seen as smart and similar to companion animals such as dogs (Kupsala et al., 2016), there is also a high level of hostility toward wolves (due to livestock predation and perceived threat to humans) and criticism of current wolf conservation practices, especially in rural areas of Finland (Komi & Kröger, 2023). Nevertheless, as the threat of wind energy develops, wolves are at times viewed locally as companions in activism against wind power infrastructure. Wind turbines are viewed as threatening human and animal activities simultaneously. As a result, human care for endangered animals may become more pronounced, as they are viewed as the victims of the rapidly spreading technologies. Appeals to securing animal well-being also make local claims against wind power development stronger, and therefore such themes often appear in Facebook groups' discussions. For example, in

[8] YLE News Finland, May 11, 2023: https://yle.fi/a/74-20031079

[9] "Susi on yhä useammin tuulimyllyn tiellä." Source: Kaleva Media, https://www.kaleva.fi/sudet-hidastavat-tuulivoimahankkeita-utajarvella-s/5165375 (accessed on February 20, 2024)

[10] "Täydellinen yllätys: Sudet kaatoivat Kajaanin tuulivoimahankkeen." Source: Maaseudun Tulevaisuus, https://www.maaseuduntulevaisuus.fi/uutiset/793dff65-b43e-5f22-861b-fff-d55485ea8 (accessed on February 28, 2024).

2024 the *Tuulivoimalapolitiikka* group discussed the Finnish Environmental Institute (LUKE) study on the effect of wind power on wolves. While some commenters noted that it would be "half-good" if wolves were chased away by wind power developments, others expressed frustration over the wolves' habitat was being ruined by the same development. This short discussion thread represents the complexity of attitudes toward the more-than-human impacts of wind energy.

These examples illustrate how wind energy infrastructure is often portrayed on social media and in newspaper publications as something that disrupts human relationships with animals, plants, and landscapes. The photos posted in Facebook groups often show animals appearing disturbed or even injured by wind turbines. These structures are depicted as imposing, dominating the surrounding environment. Discussions on social media frequently describe wind energy infrastructure as a threat to both human and animal well-being. As a result, the relationships between humans and other species are reconsidered in this context, where both face similar threats. The next section of the chapter will further explore how wind energy development can lead to a rethinking and transformation of previously established connections between humans and non-human actors.

Wind Energy Embedded in More-Than-Human Connections

As Paul Gipe (2018, p. 4) notes, the naming of wind infrastructural elements initially posed a dilemma. On the one hand, the term *wind power plant* would bring attention to the technological success of the wind industry and its place among other sources of electricity. On the other hand, the term *wind farm* would help to maintain an association with land stewardship and its responsible management. Other associations with agricultural production resulted in the relative popularity of the wind farm concept: for example, wind turbines were usually installed in rows that resembled crop fields. Additionally, the term *wind park* gained prominence and is today prevalently used. Gipe suggests that the proponents of the latter term insist upon it because of its positive connotations of natural reserves protected from industrial use, although wind turbine assemblages in no way resemble parks (Gipe, 2018, p. 4). This search for a suitable term demonstrates how wind infrastructure is presented both as technological innovations and as an integral part of the natural landscape in close entanglement with other actors in the environment.

In the previous section, I discussed how wind infrastructure is repeatedly portrayed in Facebook thematic groups through opposition metaphors or through words like "ruining" or "damaging" when referring to the landscape's aesthetic appeal or animal well-being. Interestingly, despite the focus on opposition and conflict, wind turbines are simultaneously imagined through established connections between humans and more-than-human actors in the landscape. One of the examples both in social media and newspaper publications is the comparison between wind turbines and alien plant species. In 2022, a post appeared in the *Tuulivoimalapolitiikka* group with several photos of wind turbines alongside pictures of the lupine plant (*Lupinus*). In the accompanying message, the user expressed regret that windmills do not blend into the peaceful Finnish landscape. The user went on to compare wind turbines to the invasive lupine plants, noting that, like the plants, they cannot be removed and continue to spread. Wind turbines were also compared to alien or invasive species (in Finnish, *vieraslajit*) in an opinion piece published in the *Valkeakosken Sanomat* newspaper in December 2019. While the invasive lupine plant was not mentioned there specifically, the writer noted that "the Windmill is an alien species in the Finnish nature."[11]

In an article published by Finnish media portal *Yle* in 2015, herbaceous lupine (*Lupinus polyphyllus*) is named among the five invasive species that pose a threat to Finnish nature.[12] It covers large areas close to roads, pastures, and riverbeds, and due to the alkaloids it contains, the plant may be poisonous to certain species, for example, butterflies. At the same time, as the article in *Yle* also stated, there has been a debate considering the role of lupine, and there are opinions that it does not have a pronounced negative impact on its surroundings. Still, lupine is often treated negatively; for example, in Icelandic public discourse, it has been seen as a threat to the purity and beauty of Icelandic nature (Kuprian, 2018). The rapid expansion of lupine in Iceland generated worries among conservationists and ecologists, and currently, the debate over its management is widespread (Hautala & Kulmala, 2023). In Finland, a recent study of a Finnish-language discussion forum demonstrated that the word "lupine" (in Finnish, *lupiini*) is the most popular reference when speaking about invasive species; in such discussions, lupine was blamed for suppressing heritage Finnish species from roadsides and fields (Uusitalo & Suomalainen, 2023).

Attitudes to invasive species vary. The crucial factor is their perceived ecological and economic threat, as well as the level of awareness about their possible negative impacts. Therefore, their "non-nativeness" per se may not be seen as problematic until they are directly associated with ecological or economic impacts such as out-competing native plants and causing ecosystem changes, affecting farmland, or causing fires (Lovelock et al., 2022). At times, however, the fight for the "purity" of nature may take radical forms. Thus, Paolo Bocci (2017) reflects on an example of goat eradication on the Galapagos islands carried out from 2001 to 2006, when care for endangered tortoises took the form of mass, targeted killing of another species—one seen as alien. Interestingly, there is evidence that offshore wind turbines may influence the

[11] "Tuulimylly on vieraslaji Suomen luonnossa." *Valkeakosken Sanomat*, December 2019. Source: https://www.valkeakoskensanomat.fi/puheenvuoro/art-2000006979241.html

[12] YLE, "Top five invasive species that pose a threat to Finnish nature," July 10, 2015: https://yle.fi/a/3-8146734

spread of non-native species.[13] Therefore, the negative public perceptions of wind infrastructure and invasive species may reinforce each other in the future.

When speaking about the relatively new phenomenon of wind turbines, the affected residents refer to those forms of interaction with the landscape that are more familiar to them, such as the negative public perceptions of invasive plants. Similar to the debates over the role of lupine, wind turbines are either viewed as positive additions to the landscape or as alien species spoiling its natural beauty and leading to ecosystem changes. At the same time, the perceptions of a certain species' nativity are fluid. The animals affected by the "invasive" wind turbines, such as wolves or white-tailed deer, may also be viewed as alien species to Finnish nature in other contexts and situations. The attitudes to such animals also vary from rather positive or neutral to openly negative. However, when confronted with a new perceived threat, wolves and deer form unexpected alliances with humans, now seen not as alien forces, but as companions in the struggle against the spreading technologies.

These examples demonstrate how renewable energy infrastructure becomes embedded in the already existing relations between humans and more-than-human actors. As the new technologies become a part of more-than-human entanglements, they may add new angles and shades to these connections, deepening, strengthening, or weakening them. The development of affordable and clean energy sources may also lead to the formation of completely new and at times unexpected links and alliances between humans and more-than-human actors in a common resourcescape. Acknowledging these connections would mean situating and discussing wind energy development as a part of local histories of resource use.

Conclusion

Inspired by the powerful *Oikos* performance at the Helsinki Biennial 2023, this chapter began with the question of whether the coexistence of renewable energy infrastructure, humans, and other living beings is possible, and what is gained or lost when such coexistence is attempted. By analyzing how the achievement of SDG 7 "Affordable and clean energy" in Finland interferes with the established connections between humans and more-than-human actors, this chapter brought forward the notion of resourcescape. In using this notion, I aim firstly to demonstrate that the various modes of extraction in a landscape do not happen separately, but are connected through common histories, shared local perceptions, attitudes, or practices. Additionally, this notion brings attention to the ways landscapes become resourceful. They may be valued as aesthetically pleasing surroundings, as hunting or fishing areas, as places of shared histories, or as extractive sites. These multiple aspects of landscape resourcefulness are interconnected and may complement or limit one another.

[13] Tethys Short Science Summary: https://tethys.pnnl.gov/summaries/short-science-summary-offshore-wind-farms-stepping-stones-non-indigenous-species

This chapter highlights the discrepancy between the public promotional images of wind energy infrastructure and the way residents portray them in social media groups for communication and information sharing. In promotional images, wind turbines are often shown from above, set against an "empty" landscape free of human or animal activity. Alternatively, turbines may be depicted alongside animals, suggesting a natural coexistence with emerging technologies. In contrast, social media groups typically show wind turbines from below, making them appear imposing and dominant over the landscape. When a wind turbine is pictured near animals, it is often shown as disruptive or harmful. Through these images and related comments, users convey that interactions between wind energy and the surrounding landscape are complex and often negatively affect wildlife and human life.

At the same time, while this rather direct message is conveyed, there are also less direct implications following from the posts and pictures shared by social media users, as well as local newspaper publications. The examples of golden eagles and wolves potentially damaged by wind power construction show that opposition toward renewable energy may stimulate increased attention toward vulnerable or endangered species, including ones traditionally seen as controversial. Thus, local opposition toward wind power construction may lead to the formation of new alliances between humans and more-than-human actors.

In addition, when trying to voice their concerns related to the rapid expansion of wind power, the affected residents turn to already established and familiar connections with the landscape, finding parallels between wind turbines and invasive alien species such as the lupine plant. So while wind energy is often pictured as a new phenomenon, it nevertheless becomes a part of the habitual local ways of interacting with the landscape and its resources. These previous interactions may inform local views regarding renewable energy development.

When speaking about the development of affordable and clean energy, it is crucial not to refer to renewable energy as a standalone phenomenon, but instead to situate it among other modes of resource extraction and more-than-human interactions. Paying attention to the materiality of wind energy infrastructure as a component of a complex resourcescape will potentially lead to a more nuanced understanding of its interrelations with mammals, birds, insects, plants, and human beings, not simply as an alien force but as a newer component in established entanglements. This analysis demonstrates how the achievement of SDG 7 could potentially alter human connections with non-human actors. It is important to pay attention to these changes in the framework of SDG discussions in Finland, as neglecting them would mean deepening the inequalities between regions and alienating rural, northern, and Indigenous residents.

References

Appadurai, A. (1996). *Modernity al large: Cultural dimensions of globalization* (Vol. 1). University of Minnesota Press.

Berling, T. V., & Surwillo, I. (2022). Facing the double crisis: Geopolitics and climate change put a strain on the Baltic Sea region energy security. *Trans European Policy Studies Association, 5*.

Bocci, P. (2017). Tangles of care: Killing goats to save tortoises on the Galápagos Islands. *Cultural Anthropology, 32*(3), 424–449.

Choi, B. (2022). Challenging wind power as a source of wealth in Penghu, Taiwan: Toward an anthropology of wind's entangling materiality. *Asian Anthropology, 21*(2), 81–99.

Dunlap, A., & Marin, D. (2022). Comparing coal and 'transition materials'? Overlooking complexity, flattening reality and ignoring capitalism. *Energy Research & Social Science, 89*, 102531. https://doi.org/10.1016/j.erss.2022.102531

Ey, M., & Sherval, M. (2016). Exploring the minescape: Engaging with the complexity of the extractive sector. *Area, 48*(2), 176–182.

Ferry, E., & Limbert, M. (2008). Introduction. In E. Ferry & M. Limbert (Eds.), *Timely assets: The politics of resources and their temporalities* (1st ed., pp. 3–24). School of Advanced Research Press.

Figari, H., Leiren, M. D., & Krange, O. (2024). After the battle: Emergent norms and the silencing of dissent in a Norwegian wind power community. *Energy Research & Social Science, 118*, 103765.

Gipe, P. (2018). *Wind energy for the rest of us: A comprehensive guide to wind power and how to use it*. Wind-works.org.

Günel, G. (2022). Accumulation: Exploring the materiality of energy infrastructure: Energy. In M. Hojer Bruun et al. (Eds.) The Palgrave Handbook of the Anthropology of Technology (pp. 689-702). Singapore: Springer Nature Singapore.

Hautala, R., & Kulmala, J. (2023). How erosion and other natural forces have changed the Icelandic landscape—is Nootka lupine fixing any of these problems? (Luonnon muovaama Islanti: Alaskanlupiinin (Lupinus nootkatensis) kiistanalaisuus maisemassa ja ekosysteemissä). *Suoseura—Finnish Peatland Society, 74*(1–2), 119–127.

Janhunen, S. (2018). Determinants of the local acceptability of wind power in Finland. PhD Dissertation. Lappeenranta University of Technology.

Janhunen, S., Hujala, M., & Pätäri, S. (2014). Owners of second homes, locals and their attitudes towards future rural wind farm. *Energy Policy, 73*, 450–460.

Johansen, K. (2021). Blowing in the wind: A brief history of wind energy and wind power technologies in Denmark. *Energy Policy, 152*, 112139.

Kaisti, H., & Käkönen, M. (2012). Actors, interests and forces shaping the energyscape of the Mekong region. In Forum for Development Studies (Vol. 39, No. 2, pp. 147–158). Routledge.

Kaldellis, J. K., & Zafirakis, D. (2011). The wind energy (r) evolution: A short review of a long history. *Renewable Energy, 36*(7), 1887–1901.

Keeling, L., Tunón, H., Olmos Antillón, G., Berg, C., Jones, M., Stuardo, L., et al. (2019). Animal welfare and the United Nations sustainable development goals. *Frontiers in Veterinary Science, 6*, 336.

Komi, S., & Kröger, M. (2023). Predators in the web of life: World ecology of historical human–wolf relations in Finland. *Journal of Agrarian Change, 23*(3), 500–517.

Kopnina, H. (2020). Anthropocentrism: Problem of human-centered ethics in sustainable development goals. In *Life on land* (pp. 48–57). Springer International Publishing.

Kuprian, A. (2018). Negotiating contested landscapes: The lupin controversy in Iceland. *Northern studies, 49*, 25–37.

Kupsala, S., Vinnari, M., Jokinen, P., & Räsänen, P. (2016). Public perceptions of mental capacities of nonhuman animals: Finnish population survey. *Society & Animals, 24*(5), 445–466.

Lempinen, H. (2018). At the margins of the Barents energyscape. *Barents Studies: At the Economic, Social and Political Margins, 5*(1), 13–35.

Lovelock, B., Ji, Y., Carr, A., & Blye, C. J. (2022). Should tourists care more about invasive species? International and domestic visitors' perceptions of invasive plants and their control in New Zealand. *Biological Invasions, 24*(12), 3905–3918.

Össbo, Å. (2023). Back to square one. Green sacrifice zones in Sápmi and Swedish policy responses to energy emergencies. *Arctic Review on Law and Politics, 14*, 112–134.

Poelzer, G., Linde, S., Jagers, S. C., & Matti, S. (2021). Digging in the dark: Reviewing international literature to address impending policy challenges for Swedish and Finnish mining. *Mineral Economics, 34*, 225–238.

Renewable Energy Projects on Sámi Territory. (2022). Fact sheet by the Sámi council. https://lcipp.unfccc.int/sites/default/files/2022-06/Factsheet%20-%20Renewable%20Energy%20Projects%20on%20Sa%CC%81mi%20Territory.pdf

Retter, G.-B. (2021, November 25). Indigenous cultures must not be forced to bear the brunt of global climate adaptation. *Arctic Today.* https://www.arctictoday.com/indigenous-cultures-must-not-be-forced-to-bear-the-brunt-of-global-climate-adaptation/

Salonen, H., & Tomrén, L. (2023). Can local value creation induce a sense of justice during green transitions?: A study of six rural areas in Denmark, Finland, and Norway. https://pub.nordregio.org/r-2023-9-can-local-value-creation-induce-a-sense-of-justice-during-green-transitions/index.html

Seipiharju, S., & Oleander-Turja, J. (2023, August 24). *Vihreä Siirtymä nähdään Haasteena Saamelaisnuorten Tulevaisuudelle, Pelkona Kulttuurin Ja ympäristön Tuhoutuminen: "sukupolvemme kärsii jo."* Yle Uutiset. https://yle.fi/a/74-20046532

Taylor, N., Valencia-García, L. D., VandenBroek, A., Stinnett, A., & Allen, A. (2023). Ethics and images in social media research. First Monday. https://firstmonday.org/ojs/index.php/fm/article/view/12680/10940

Toivanen, R., & Cambou, D. (2021). Human rights. In K. Parker & R. Toivanen (Eds.), *Situating sustainability: A handbook of contexts and concepts* (pp. 51–62). Helsinki University Press.

Torpman, O., & Röcklinsberg, H. (2021). Reinterpreting the SDGs: Taking animals into direct consideration. *Sustainability, 13*(2), 843.

Townsend, L. & Wallace, C. (2016) Social media research: A guide to ethics. https://www.gla.ac.uk/media/Media_487729_smxx.pdf

Tsing, A. (2005). *Friction: An ethnography of global connection.* Princeton University Press.

Tsing, A., & Lassila, M. (2017). Interview with Anna Tsing. *Suomen Antropologi: Journal of the Finnish Anthropological Society, 42*(1), 22–30.

United Nations. (2015). Transforming our world: The 2030 Agenda for Sustainable Development. Available at: https://sdgs.un.org/2030agenda

Uusitalo, H., & Suomalainen, K. (2023). Ecolinguistic approach to online Finnish discourse on invasive alien species. *Language@ Internet, 21*, 57–78. https://www.languageatinternet.org/articles/2023/uusitalo

Varfolomeeva, A. (2023). Indigenous connections with the resourcescape in the Russian north and Siberia. *Globalizations*. https://doi.org/10.1080/14747731.2023.2171221

Vincent, W. F. (2020). Arctic climate change: Local impacts, global consequences, and policy implications. In K. Coates & C. Holroyd (Eds.), *The Palgrave handbook of Arctic policy and politics* (pp. 507–526). Palgrave Macmillan.

Wolf, S., Teitge, J., Mielke, J., Schütze, F., & Jaeger, C. (2021). The European Green Deal—More than climate neutrality. *Intereconomics, 56*(2), 99–107.

Zilliacus, J. (2022, March 30). Green coloniality of power over Sámi in Finland? Social Exclusion: Blog of Åbo Akademi. https://blogs.abo.fi/socialexclusion/2022/03/30/green-coloniality-of-power-over-sami-in-finland/

CHAPTER 14

The Drama of the Missing Commons in the SDGs

Tobias Haller and Ilkhom Soliev

Introduction: The Missing Commons in the SDGs

The Sustainable Development Goals (SDGs) were created by the United Nations (UN) to go beyond the Millennium Development Goals (MDGs) and address the issue of sustainability. The 17 goals and their sub-goals contain various recommendations and discuss key environmental and human rights issues, but with a focus on implementation by national governments and the private sector. The question of how local societies and groups—especially in the Global South—can sustainably manage their collectively owned common-pool resources (CPRs such as fisheries, wildlife, pastures, forests, water, etc.) is not a topic in the SDGs. Although the issues of resource governance are addressed in the Brundtland Report and in the MDGs, as well as in the SDGs, they never mention widespread existence of systems of collective ownership (commons property) of natural resources and the ability of local groups to manage them by means of locally crafted rules and regulations, so-called institutions. Although some of the globally pressing social issues such as poverty and justice are comprehensively included in the SDGs, a closer look at the goals

T. Haller (✉)
Institute of Social Anthropology, University of Bern, Bern, Switzerland
e-mail: tobias.haller@unibe.ch

I. Soliev
Department of Environmental Sociology, Institute of Sociology, Martin Luther University Halle-Wittenberg (MLU), Halle (Saale), Germany

German Centre for Integrative Biodiversity Research (iDiv) Halle-Jena-Leipzig, Leipzig, Germany
e-mail: ilkhom.soliev@soziologie.uni-halle.de; ilkhom.soliev@idiv.de

R. Toivanen et al. (eds.), *Decolonizing the Sustainable Development Goals*,
https://doi.org/10.1007/978-3-032-14263-4_14

reveals problematic omissions in relation to other key social issues that are at least as important for tackling some of the current environmental and development crises. In particular, the commons, once considered a major issue in the context of sustainability and reflected in Elinor Ostrom's Nobel Prize (Ostrom, 1990), are completely left out. Ostrom's work was based on solid scientific evidence that local communities were often able to establish institutional arrangements for the sustainable management of common-pool resources (CPRs) associated with the land (pasture, wildlife, fisheries, forestry, water, etc.). It also drew, in particular, on European empirical examples, especially the work of the US social anthropologist Robert Netting (1981), who had researched such regulations in the community of Törbel in the Upper Valais. While the sustainable management of common-pool resources is directly linked to SDGs 14 and 15, CPRs are also relevant to all other SDGs (see, e.g. hunger, energy, consumption, climate change and biodiversity loss). 'Commoning' as a process of shared governance also has a lot to do with SDGs 16 (peace, justice and strong institutions) and 17 (participation).

How Important Are Commons for the SDGs, and What Are the Specific Policy Requirements?

The relevance of the commons—the management of CPRs by common property institutions—especially regarding the sustainable use of resources as described in the SDGs, is still of great global relevance as we will show. *First*, in the Global North, the value of CPRs has been massively reduced by structural change (industrialization and wage labour), and the preservation and provision of ecosystem services through the maintenance of cultural landscapes is endangered but not (yet) leading to immediate livelihood issues. However, as the example of Switzerland shows, it is mainly local commoners' organizations, whose land rights are secured, and who are capable of maintaining these collectively owned landscapes and their biodiversity. They do this despite being economically not profitable because they have an identity value and provide an SDGs service without recognition (see Haller et al., 2021a, b). In the Global South, however, where local common property is not secured but under pressure from land grabbing and land exploitation, and where at the same time wage labour is often not possible, the commons continue to be one of the few guarantors of resilient survival under threat (see Giger et al., 2019; Kronenburg et al., 2022; Haller, 2022).

Second, the absence of the commons in the SDGs is all the more tragic because not only are they not recognized, but there is also a danger that states and the private sector will be able to further expand their interests and engage in commons grabbing for conservation, resource extraction or expanding renewable energy production legitimized in the name of the SDGs (see Larsen et al., 2022). Overview studies globally show how important the commons would be for biodiversity, climate protection, water governance and renewable energy provision. In Switzerland, as a European case, for example, collectively owned pastures and forests as well as water sources are found in many cantons and have been managed for centuries, in some areas even making up to 70% of the territory (Haller et al., eds., 2021a, b). Studies on farmers and pastoralists in the Global South and North likewise indicate that most of them rely on the commons (Haller, 2022; Giger et al., 2019; Soliev & Theesfeld, 2020; Kronenburg et al., 2022). Therefore, it would now be important to insist on collective resource rights in the context of sustainability in national implementations and to influence national implementation policies to this end (Haller et al., eds., 2021a, b).

Why are SDGs relevant to the debate on the commons in practice? *Firstly*, the SDGs cover virtually all areas of action in a society—social, economic and environmental—as this international framework brings together development and environmental issues and expands them to include the issues of achieving and sustaining greater justice and peace. *Secondly*, the SDGs represent a serious global commitment in terms of their scope, but also in terms of financial resources, as it is estimated that USD 4.2 trillion per year—by comparison, Germany's GDP is around USD 3.8 trillion—is to be made available for the implementation of the SDGs (Joint SDG Fund, 2022). At the same time, the SDGs provide political resources based on a hegemonic discourse that prescribes sustainable development by states and the private sector, as Larsen et al. emphasize (2022). The authors show that while SDGs directly target commons-relevant resources and institutional arrangements on a large scale, the dominant actors with their environmental discourse have so far not left much room to include the commons in the global sustainability debate. The omission of the commons in the SDGs will have two effects: *Firstly*, state elites can continue to legitimize the appropriation of commons, as commons regimes as well as common property are not officially recognized in the SDGs. *Secondly*, the exclusion of the commons reduces the negotiating power of local collective owners, which could be used to claim better participation in the definition and implementation of the SDGs. Therefore, we also see two corresponding policy needs to: (a) ensure that SDGs do not become an acronym for 'Sanctioning Disciplined Grabs' (see also the title of Larsen et al., 2022) and (b) ensure that local owners of collective property (pastures, forests, water resources, hunting grounds, etc.) can use and develop alternative and/or complementary mechanisms to state-led and private property-based governance during the implementation phases of the SDGs. This leads us back to critically discuss the definition and overuse narratives of the commons.

Definition of the Term 'Commons', Narratives and Implications

A first look at previous major international development and environmental policy agreements shows that references to local collective resource ownership are often missing or only marginally integrated. This raises questions with regard to the definition of the English term 'commons'. In the academic debate before and after the seminal work of Elinor Ostrom (notably Ostrom, 1990), the term 'commons' is used more broadly, and analysing what is understood by the term and how it relates to the SDGs can already shed light on its early beginnings, its career and its subsequent disappearance, at least in policy texts. Our literature review shows that numerous terms were used in academic discourse between the 1980s and 2010s: 'commons', 'governance of the commons', 'common property resources (CPR)', 'common property institutions', 'common pool resources', 'common property regimes', 'open CPR', 'closed CPR', 'commonly held resources', 'commoner', 'commoning' (e.g. Ostrom, 1990; Berkes, 1999, 2009; Sarker & Blomquist, 2019; Quintana & Campbell, 2019; Haller et al., 2019; Cole, 2020). In fact, the definitions and perceptions of the term have been interpreted so broadly that Sarker and Blomquist (2019), for example, even speak in the title of their paper of 'misconceptions of commons'. Indeed, since the misdescription of the term 'commons' as equating to open access in Hardin's (1968) article, there has been debates to gain a more accurate and empirically grounded understanding of the commons. The view that commons must necessarily lead to the overuse of resources due to open access and that state ownership and privatization are sustainable alternatives can be refuted as follows.

First, however, this requires a distinction to be made between common-pool resources (CPRs) and common property (see also above): CPRs are regenerable resources with specific characteristics that make the exclusion of users of these resource systems difficult, but not impossible in groups (excludability, e.g. a large pasture, fishing area or forests).On the other hand, CPRs are 'subtractable', i.e. a utilized unit is no longer available and must regenerate itself. However, this only refers to the quality of a resource, which does not necessarily determine the ownership system and governance. This is because CPRs can be in an open access system, controlled or privatized by the state (state or private property) or owned by local communities and managed as common property (including through membership and self-organized governance institutions, see also Ostrom, 1990, for this distinction). While open access can lead to overexploitation, there is extensive empirical evidence worldwide that open-access and commons cannot be equated and management of the commons (a) is based on local community ownership and (b) is the basis of locally agreed institutions (rules).

Ostrom concluded in her work that if local community ownership institutions are robust, they can be a promising way forward for the sustainable use of natural resources. Indeed, the empirical basis of this discussion, often based on social anthropological research, formed the basis for Ostrom's Nobel Prize (see Netting, 1981; Ostrom, 1990). It represents an important contribution to the sustainability debate. Such an understanding of the commons allows us to examine whether local commons are recognized as common property regimes in sustainability policy and not just misunderstood as freely accessible or ambiguous 'global' commons. Our discourse analysis shows that local commons are clearly absent from the SDG framework texts, such as the 1987 Brundtland Report 'Our Common Future', which mentioned commons exclusively in terms of global commons, while emphasizing their importance for the collective future. However, the report, which formed the basis for the discourse on sustainable development, did not focus on local collective management of common resources, which would otherwise also recognize local communal property rights and the right to set rules and regulations. The focus is rather on global governance systems. This becomes clear in formulations such as '[…] "the global commons" - those parts of the planet that fall outside national jurisdictions' (WCED, 1987), which are then illustrated with the examples of 'oceans', 'outer space' and 'Antarctica'. Some features of the commons were then also mentioned rather superficially in the formulation of 'shared ecosystems', coupled with a no less unclear formulation of 'development of shared ecosystems'. The term 'development' is used in relation to the term 'ecosystems' without clarifying what this entails. In any case, the historically evolved collective ownership and management of sub-elements of these ecosystems was not addressed.

Similar trends, with some differences, could then be observed in the subsequent influential international documents that shaped the sustainable development agenda. An example is the Millennium Development Goals of 2000—probably the most influential development agenda before the SDGs. There are very weak and only implicit references to the characteristics of commons as a common resource system, while the recognition of common property regimes is completely absent, as is the mention of local institutional resource governance. This is evident in phrases such as 'ensure environmental sustainability', 'reverse the loss of environmental resources' and 'reduce biodiversity loss', all of which are partly implicitly treated as systems of shared value and require collective action, but do not explicitly integrate mentions of the commons. Finally, there are still very weak and similarly only implicit references to the commons in the SDGs but only related to equality or equity of sharing and ownership in general: 'All men and women […] have equal rights to 'economic resources', 'equal access to', 'ownership and control over land and...'. However, it could be argued that this is actually a step backwards for the visibility of collective ownership systems compared to the times of the Brundtland Report. Not only is the term 'commons' not mentioned, but there are no considerations of local forms of governance other than governance by the state and the private sector. The SDGs give the impression that the latter are seen as the only forms of governance capable of reversing the social and environmental crises of our time and bringing about comprehensive sustainability.

What Could Explain the Absence of the Commons in the SDGs?

How can we explain this puzzling absence of the commons in the SDGs, despite the growing importance of the commons discourse, reinforced in particular by Ostrom's Nobel Prize, and a wide range of academic debates that refer to the commons as an important pathway to sustainability? One reason could lie in the definitional challenges discussed above. The term 'commons' requires a nuanced understanding. Although the academic literature discusses and reflects this (Acheson, 1989; Ostrom, 1990; Feeny et al., 1990; Berkes, 1999; Sarker & Blomquist, 2019; Quintana & Campbell, 2019; Haller et al., 2019; Cole, 2020), the term is very rarely used in policy and practice in the field of sustainable development and, if at all, is often used implicitly and in very broad terms.

Another important explanation lies in the desire of states and governments to reduce the voice of other actors, and by not taking common property into account, this can be made easier, as states with their governments and the private sector appear to be the only legitimate stakeholders in the development and implementation of the SDGs (Haller et al., 2018; Larsen et al., 2022; Haller & Lamatsch, 2023). This lack of participation by actors other than state and private actors serves the self-interests of powerful interest groups and the dominance of capitalist economic forms. This could also be the reason why the concept of common goods is viewed with suspicion. The prevailing narrative of commons as systems of open access emphasized by governments and private actors is thus also conducive to not taking self-management alternatives seriously or even interpreting them as a kind of rejection of capitalist state and privatization solutions. Not only is this a widespread misunderstanding (Sarker & Blomquist, 2019), but it is also not apparent that Ostrom's analysis of the commons was determined by such an explicit political anti-state or anti-private property view. Nor does it correspond empirically to the functioning of many common property regimes, especially in Europe and Switzerland, where local concepts of household ownership of fields and meadows were combined with access to communal forests, pastures and water (see also Haller et al., 2021a, b). However, a critical point on Ostrom's analysis is that external and internal power relations that characterize common property institutions were not at the centre of her work and, thus, important dynamics in the design of rules and regulations for collective resources are generally little studied in the commons debate (see Haller et al., 2021a, b). Apart from this criticism towards the commons scholarship that largely remained void of power analysis, the anti-state and anti-capitalist connotation wrongly or strategically attributed by governments and the private sector to the commons approach could explain to some extent why the commons are absent from the SDGs documents.

Another aspect that could explain the conundrum around the omission of commons in the SDGs is the lack of recognition in the scientific literature, which since the 1980s has emphasized that what is called 'nature' is in fact the result of human landscape change leading to cultural landscape ecosystems (see Ellen, 1982; Fairhead & Leach, 1996; Berkes, 1999; Haller et al., 2022; Haller, 2019a). As a diverse literature from economic and ecological anthropology demonstrates, this 'nature' has been shaped through diversified, extensive and intensive economic systems of subsistence use (gathering, hunting, grazing and traditional agriculture largely for subsistence) in both the Global North and South in pre-industrial and pre-colonial times, mostly through communal resource regimes. These created what can be described as highly diverse 'cultural landscape ecosystems' around the world (see Ellen, 1982; Fairhead & Leach, 1996; Haller, 2019a, 2019b, 2022). If the SDGs are about the protection of 'nature', it can certainly be argued that 'nature' in this pure form does not exist, but that we are dealing with an immense diversity of historically evolved cultural landscapes that should be protected. This would include the commons from a different perspective, and, in this sense, the commons should actually be emphasized as the dominant form of resource ownership and governance of CPRs to be protected in the SDGs.

However, this is not the case, and behind this lack of recognition are, among other things, power-specific dynamics that prevent the restoration of the commons. This is because processes of resource curtailment, especially on formerly communal land, have not ceased with the end of colonization and the formation of independent states. On the contrary, they have tended to increase in the countries of the Global South as well as in the North, as a form of takeover processes through privatization with the support of neoliberal movements (Dell'Angelo et al., 2017; Giger et al., 2019; Haller 2019a, 2019b, 2022). The mention of the commons could lead to local demands that challenge this process (Larsen et al., 2022) and allow the 'drama of the grabbed commons' (title of paper from Gerber & Haller, 2021) to be transformed into a process of recognizing land and land-related CPRs as common property and measures to strengthen them, making them no longer readily accessible to capitalist interests. It cannot be ruled out that this is partly the reason why, despite an increasing involvement of NGOs in debates on sustainable development, commons have acquired a negative connotation from the perspective of governments and powerful actors with vested interests. We can also observe that it is only since the middle of the last decade that the term commons had acquired a much more positive connotation. This could initially be attributed to academic activists who view commons as the process of developing an alternative type of sustainability and way of life (see Helfrich & Bollier, 2015). However, as the question of power and resource rights is not raised here, such approaches are entirely compatible with neoliberal movements, governments and international development planners. How can this paradoxical change be explained? There is sufficient evidence to suggest that governments might reconsider commons as a form of self-governance, not in terms of returning property rights and benefits, but rather to reduce governance costs for the state—similar to the colonial form of indirect rule, but without the rights granted and guaranteed by common property (see also Schläppi, 2018; Haller et al., 2019).

Why Is the Omission of the Commons from SDGs Highly Problematic?

However, the SDGs were written, designed and negotiated by states even before the period of the aforementioned debate. But even the turning point described above does not yet allow for a broader take-up of the topic. While on the one hand there is still no option for recognition, on the other hand, the absence of the term 'commons' in the SDGs has problematic effects, since this absence also leaves related terms such as 'common-pool resources' and 'common property' regimes as well as locally developed 'common property' institutions likewise unmentioned in the SDGs. However, the problem is not only the lack of recognition. The problem is that the lack of emphasis on common goods in the written SDG documents further empowers governments and the private sector to use mainstream development and conservation programmes and policies for their own interests and to legitimize them with seemingly ecologically sound justifications, without truly involving and recognizing local stakeholders and their collective land and resource rights.

The SDGs therefore not only fail to represent a paradigm shift towards sustainability, but they also risk further deteriorating the status quo. The SDGs set out a broader technocratic agenda that risks reproducing the usual notions of development and protection of 'pure nature' from top to bottom, rather than from the perspective that we are dealing with cultural landscape ecosystems almost everywhere on the planet. The power of definition of what 'nature' and 'sustainability' mean thus lies not with local communities, but rather with state and private sector actors in the Global North and governments in the Global South. There are many similarities to older debates on development, such as the development criticism formulated in the 1990s by Arturo Escobar in his work *Encountering Development* (Escobar, 1994) or James Ferguson's concept of the anti-political machine, which conceals power asymmetries and state control behind ideas of development for all (Ferguson, 1994; see also Haller et al., 2018; Haller & Lamatsch, 2023). With the SDGs, there is now a danger that they have developed or are being developed into a modern green anti-political machine that obscures the power asymmetries underlying these developments. The SDGs do this by proposing green development, but at the same time concealing which powerful actors have driven and continue to drive overuse and pollution, and that the majority of global actors are being left out. Phrases such as 'leave no one behind' in the SDGs conceal the structural processes that lead to many people on the planet being left behind, because the reasons for leaving them behind are not questioned and their abolition is not demanded. Political ecology scholars argue that neoliberal exploitation relies on the continuous marginalization of the commons, as they provide a cheap and essential resource. This includes both the labour that depends on the commons for survival and the common-pool resources (CPRs) that sustain collective groups, which, in turn, reproduce these systems at minimal cost to capitalism (see De Angelis, 2000, 2023).This becomes clear, for example, in the debate on nature conservation in Africa. Here, the protection of pure nature in the context of the SDGs not only obscures the fact that these are cultural landscapes that are the result of earlier common property systems with their institutions, but also that this protection generates a high level of tourism. This must also be protected, but to the exclusion of local farming and nomadic populations, whose use of the land, based on common property institutions, has created the diverse grassland areas and habitats for wild animals and plants in the first place. Instead, local nomadic cattle herding groups, for example, are now being blamed for overgrazing and seen as the threat to wild animals and predators in these ecosystems, which leads to exclusion and commons grabbing (see a wide range of literature on this topic, e.g. Neumann, 1998; Brockington, 2002; Galvin & Haller, eds., 2008; Brockington et al., 2008; Hall et al., 2015; Haller, 2022). The latest debate about commoners in Tanzania's Ngorongoro Crater illustrates this situation (see Human Rights Watch report in 2023).[1] In addition, both older and more recent research shows that it is not only about commons grabbing but also often about human rights violations, because state and semi-state wildlife protection organizations often use violence against 'poachers' from the local population, often with fatal consequences (see Haller, 2013; Ramutsindela et al., 2022). Not only expulsion and human rights violations, but also the financial and workload-related pressures of environmental protection requirements (from biodiversity to the protection of predators such as wolves) are top-down strategies that are also readily legitimized in Europe and Switzerland with reference to the SDGs. These strategies push European commoners to their breaking point, while their cultural landscape services are exploited for tourism and not sufficiently subsidized (see also Haller et al., 2021a, b, for Switzerland). This shows that the wording of the 17 goals and sub-goals is formulated in such a way that they largely do no harm to governments, and in this sense, there can be less talk of 'leaving no one behind' and more of 'leaving no governments behind'.

Five Concrete Threats Posed by the Formal Absence of the Commons in the SDGs

From an interdisciplinary perspective of anthropology, geography and sociology, but especially from the perspective of a new institutional political ecology, we see five main threats that the SDGs may pose to local citizens (see Haller et al., 2018; Haller, 2019a, 2019b; Larsen et al., 2022; Haller & Lamatsch, eds., 2023):

1. Disregarding the root causes of environmental degradation and acting as a green anti-politics machine help to conceal unequal development and destructive power relations in interconnected global-local ('glocal') economies.
2. Misreading cultural landscape ecosystems as 'pure nature', a colonial view of locally managed cultural landscapes since the pre-colonial past, especially in the Global South, contributes to the legitimization of the discourse on green interventions.
3. The current formulations of the SDGs contribute to the continuation of encroachments because they allow the non-recognition of the shared management of community resources, while at the same time obscuring

[1] See https://www.hrw.org/news/2023/12/01/tanzania-tour-risks-whitewashing-maasai-forced-evictions [last visited in February 21, 2025].

the dominance of state-led grabbing processes, which, after the colonial era, take the form of land grabbing as commons grabbing (land, pastures, wildlife, fisheries, forestry, water, etc., see Giger et al., 2019 with reference to the Land Matrix).

4. The lack of recognition of commons in the SDGs undermines local participation in the definition of new rules for sustainable governance, which is now largely controlled by the state, the private sector and state-recognized NGOs. These actors are often not interested in self-defined local participation and a participatory formulation of new goals and local institutions. It allows governments to exert control by labelling state-driven processes of top-down participation as commons (see also the work of Fukuda-Parr, 2019).
5. The absence of commons in the SDGs expands the options for powerful actors, who now have a well-supported environmental agenda, since the only participating actors are states, powerful environmental NGOs and the private sector (multinational corporations). This increases the risk that the SDGs could be used by governments and the private sector to appropriate more commons. This could be done, for example, through public-private partnerships (PPPs), legitimized by narratives of poverty alleviation and sustainable development, to remove property rights from local citizens, often in connection with nature reserves and green energy development projects (biofuels, solar and wind energy (see various contributions in the anthology by Kronenburg et al., 2022; Haller & Weissman, 2023; Haller & Weissman, eds., 2024)). In such projects, companies can justify their activities by developing corporate social responsibility (CSR) systems related to the SDGs and concealing exploitative labour relations (Gerber & Haller, 2021). Furthermore, governments are enabled to replace genuine participation with top-down participation and processes of suppressing criticism by including key terms. Talking about justice without meaning redistribution, emphasizing partnership while at the same time defining rules from above and criticizing local population groups while at the same time concealing the main perpetrators of environmental damage are strategies that are emerging. This shows that the SDGs fit well into the idea of a neocolonial and neoliberal global governance system.

These five critical observations (see also Larsen et al., 2022; Haller & Lamatsch, eds., 2023) can be illustrated at least in the following six empirically researched global fields of action, in which the accumulation of wealth by some through the expropriation of the majority of the population is made easier with specific SDGs for a basis of legitimacy:

- *Agro-industrial sectors* (land grabbing, deforestation, e.g. palm oil industry), see SDGs 7, 9, 13

- *Mining*, e.g. *for electric and climate-smart cars* (refers to the SDGs on green energy and climate change), SDGs 7, 13
- *Protection of biodiversity*, e.g. *forest protection* (Half Earth and 30 × 30 agendas), e.g. protection relates to SDGs 14, 15 (life in water and on land)
- *Top-down development programmes* related to hunger and development (from SDG 1 to SDG 3) (e.g. agroecological farming, aquaculture)
- *Mega-infrastructure projects* are praised for sustainable energy production (7) and infrastructure (9) and conceal the theft of common goods (pasture land, fisheries, wildlife, local agricultural land, access to water)
- *Hegemonic territorial design in government control:* Land use planning and neoliberal land reforms for continued land grabbing based on a process of territorialization and increasing state control, SDGs 7, 9, 13, 14, 15, 17

Last but not least, the SDGs can not only legitimize commons and resilience grabbing processes based on a neoliberal green agenda (see Haller et al., 2020) but also convert the legitimacy and inclusion of criticism into productive discursive assets to accumulate more capital. Options for this could include tax policies with reference to the SDGs, the use of further regional development banks and other funds (e.g. at the EU or inter-African or inter-American level), and finally the Global Funds for Sustainable Development.

Conclusions: Possible Strategies for Political Action

What political options are there in the context of this criticism? The SDGs are written and cannot be rewritten, although an experiment in this direction produced interesting aspects for constructive criticism (see Haller et al., 2018). However, in our opinion, there are two measures that we propose for action in the political sphere in the current situation:

Strategy I: Documentation of commons grabbing by the SDGs and establishment of a centralized but globally networked ombudsman's office. Further research and documentation is needed to show how the use of SDGs enables and legitimizes processes of appropriation of commons by states and the private sector (examples already documented in Kenya): The LAPPSET corridor in Kenya, which refers to the SDGs and the 2030 Agenda and thus legitimizes the undermining of fisheries and nomadic pastoralism (Werthmüller, 2024), and conservation projects such as LEWA, which are presented as a model of participation when in fact they are a form of green commons grabbing (see Weissman, 2019). These documentaries should be used to create the basis for political change that commoners would demand. Furthermore, this could be the basis for supporting local resistance and for reviewing and initiating legal action. Similar to the Land Matrix for land grabs, we could imagine a matrix for commons grabbing for SDGs.

Strategy II: Actively shaping and transforming the implementation of the SDGs in states. Continuous discourse-shaping critical dialogue and action are needed on the types of sustainability, development and goals that we want as as a society. In Switzerland, for instance, an attempt is being made to get involved in the implementation process of national SDGs in cooperation with the Swiss Association of Citizens' Municipalities and Corporations (*Schweizerischer Verband der Bürgergemeinden und Korporationen*, SVBK) and researchers from the history, geography and social anthropology departments at the University of Bern. The idea is to involve the national administrations responsible for implementation in the cooperation. In a project funded by the Swiss National Science Foundation (project Sustainable Commons Adaptations to Landscape Ecosystems in Switzerland (SCALES, see Haller et al., 2021a, b)), researchers at the University of Bern (Institute of Social Anthropology, T. Haller) show the important role of the commons for the preservation of Swiss cultural landscape ecosystems. Between 30% and 90% of all Alpine pastures and forests in the various cantons of Switzerland are collectively owned by commoners' organizations. Overall, however, there are major differences. At least one-third up to 40% of all forests and about 50% of all pastures in Switzerland are collectively owned, and, in this respect, the participation of commoners and their co-determination in the implementation of national SDG measures are more than justified. However, their situation is still fragile, as Swiss common property resources must strike a balance between low market prices for their products (agricultural dairy products and wood) and top-down planning of subsidies by the state that do not cover the costs of collective ownership for the preservation of cultural landscapes. This depends on their respective bargaining power vis-à-vis municipalities, cantons and the federal government, as well as on their economic strategies. Research has shown that innovations that help them to maintain and internally cross-subsidize market-specific parts of their property that generate a deficit increase this bargaining power. This means, for example, that income from real estate is used to cover deficits from forestry and alpine pasture use. This also helps the citizens' organizations to fulfil their sustainability goals, to be able to use their properties not for short-term financial gain but for future generations. Their valuation is not only market-oriented but also geared towards identity-forming benefits in the form of the maintenance of cultural landscape heritage. In addition, citizens' organizations are aware of the criticism of the exclusion of other users and are experimenting with inclusive internal guidelines and ways of addressing the problematic gender imbalance in their organizations that has been inherited from the past (see Haller et al., 2021a, b). There is also a good organizational basis: The more than 1600 Swiss commoners' organizations are since 1948 organized in a Swiss Commoners umbrella organization (SVBK) now trying to raise awareness among the public, but especially among political decision-makers and administrative officials, that collective resource owners must be recognized and included on an equal footing in the national implementation phase of the SDGs and in defining the post-SDG priorities. Specifically, this means recognizing the commoners' organizations and cooperatives as partners in Switzerland's written national SDGs reports to the UN and negotiating binding rules for participatory cooperation between the state and citizens' organizations in connection with the implementation of the national SDGs. This could be a first step towards a precedent of a country recognizing common property regimes for CPRs and thus common property and its relevance for the SDGs. It is hoped that the reference to state recognition in Switzerland could help civil society organizations and movements in other countries (especially local and Indigenous communities) to claim the same form of recognition.

Such steps are needed. Because despite the historical promise of the commons discourse with Ostrom's Nobel Prize seemingly influencing the early agenda for sustainable development, our analysis unfortunately shows rather negative trends. Not only have commons issues slowly disappeared from the language of the global sustainability agenda, but the latest framework of powerful SDGs provides an easier platform for the destruction and appropriation of commons. We need to better understand and document these processes and activate political opportunities to reintroduce the much-needed value of commons into the SDGs in a meaningful way, while commoners also have to organize nationally, regionally and internationally to claim back the governance of CPRs. Including them by claiming their participation in the implementation process and their ability to craft resource governance institutions from the bottom-up ('Constitutionality', see Haller et al., 2016) could be a way forward.

References

Acheson, J. M. (1989). Management of common property resources. *Economic Anthropology*, 351–378.

Berkes, F. (1999). *Sacred ecology: Traditional ecological knowledge and resource management.* Taylor and Francis.

Berkes, F. (2009). Evolution of co-management: Role of knowledge generation, bridging organizations and social learning. *Journal of Environmental Management, 90*(5), 1692–1702.

Brockington, D. (2002). *Fortress conservation. The preservation of the Mkomazi game reserve, Tanzania.* Indiana University Press.

Brockington, D., Duffy, R., & Igoe, J. (2008). *Nature unbound. Conservation, capitalism and the future of protected areas.* Earthscan.

Cole, D. H. (2020). Elinor Ostrom, Pioneer of the commons (and much more). In J. G. Laitos & J. C. Nagle (Eds.), *Pioneers of Environmental Law* (pp. 151–172). Twelve Tables Press.

De Angelis, M. (2000). Global capital, disciplinary trade and new commons. https://citeseerx.ist.psu.edu/document?repid=rep1&type=pdf&doi=8eaceec17a5dbd17d244ecc7f5ad9d79941d15a3.

De Angelis, M. (2023). The capitalist pyramid and the commons. https://europe.iasc-commons.org/iasc-europe-cis-colloquium-series2023/.

Dell'Angelo, J., D'Odorico, P., Rulli, M. C., & Marchand, P. (2017). The tragedy of the grabbed commons: Coercion and dispossession in the global land rush. *World Development, 92*, 1–12.

Ellen, R. (1982). *Environment, subsistence and system. The ecology of small-scale social formations*. Cambridge University Press.

Escobar, A. (1994). *Encountering development: The making and unmaking of the third world*. Princeton University Press.

Fairhead, J., & Leach, M. (1996). *Misreading the African landscape: Society and ecology in a Forest-savanna mosaic* (Vol. 90). Cambridge University Press.

Feeny, D., Berkes, F., McCay, B. J., & Acheson, J. M. (1990). The tragedy of the commons: Twenty-two years later. *Human Ecology, 18*, 1–19.

Ferguson, J. (1994). *The anti-politics machine. "Development" and Bureaucratic power in Lesotho*. University of Minnesota Press.

Fukuda-Parr, S. (2019). Keeping out extreme inequality from the SDG agenda–the politics of indicators. *Global Policy, 10*, 61–69.

Galvin, M., & Haller, T. (Eds.). (2008). *People, protected areas and global change, for NCCR north south, perspectives series*. University of Bern.

Gerber, J. D., & Haller, T. (2021). The drama of the grabbed commons: Anti-politics machine and local responses. *The Journal of Peasant Studies, 48*(6), 1304–1327. https://doi.org/10.1080/03066150.2020.1758673

Giger, M., Nolte, K., Ansseuw, W., Breu, T., Messerli, P., Oberlack, C., & Haller, T. (2019). Impacts of large-scale land acquisitions on common-pool resources: Evidence from the land matrix. In T. Haller, T. Breu, T. de Moor, C. Rohr, & H. Znoj (Eds.), *The commons in a Glocal world: Global connections and local responses* (pp. 257–279). Routledge.

Hall, R., Scoones, I., & Tsikata, D. (Eds.). (2015). *Africa's land rush: Rural livelihoods & agrarian change* (Vol. 36). Boydell & Brewer.

Haller, T. (2013). The Contested Floodplain. Institutional change of the commons in the Kafue Flats, Zambia. Lanham: Lexington, Rowman & Littlefield. (2013)

Haller, T. (2019a). Towards a new institutional political ecology: How to marry external effects, institutional change and the role of power and ideology in commons studies. In T. Haller, T. Breu, T. de Moor, C. Rohr, & H. Znoj (Eds.), *The commons in a Glocal world: Global connections and local responses* (pp. 90–120). Routledge.

Haller, T. (2019b). The different meanings of land in the age of neoliberalism: Theoretical reflections on commons and resilience grabbing from a social anthropological perspective. *Land, 8*(7), 104. https://doi.org/10.3390/land8070104

Haller, T. (2022). From commons to resilience grabbing: Insights from historically-oriented social anthropological research on African peasants. *Continuity and Change, 37*(1), 69–95.

Haller, T., Acciaioli, G., & Rist, S. (2016). Constitutionality: Conditions for crafting local ownership of institution-building processes. *Society & Natural Resources, 29*(1), 68–87.

Haller, T., Breu, T., de Moor, T., Rohr, C., & Znoj, H. (Eds.). (2019). *The commons in a Glocal world: Global connections and local responses*. Routledge.

Haller, T., & Lamatsch, J. (2023). New debates and implementations of the sustainable development goals (SDGs). A critical social anthropology perspective. Institute of Social Anthropology, University of Bern, Switzerland, in collaboration with students. https://2023.iasc-commons.org/wp-content/uploads/2023/06/1SubthemeSDG_workingpaper_HallerLamatsch2023.pdf

Haller, T., Liechti, K., & Mann, S. (2021a). Commons and peasant studies: Insights from social anthropology, human geography and agrarian economics. In T. Haller, K. Liechti, M. Stuber, F. X. Viallon, & R. Wunderli (Eds.), *Balancing the commons in Switzerland. Institutional Transformations and Sustainable Innovations* (pp. 45–60). Routledge.

Haller, T, Liechti, K., Stuber, M., Viallon, F. X., & Wunderli, R. (eds.). (2021b). Balancing the commons in Switzerland: Institutional transformations and sustainable innovations. Routledge. https://www.routledge.com/Balancing-the-Commons-in-Switzerland-Institutional-Transformations-and/Haller-Liechti-Stuber-Viallon-Wunderli/p/book/9780367488734

Haller, T., Ngutu, M., & Käser, F. (2020). Does commons grabbing lead to resilience grabbing? The anti-politics machine of neo-liberal agrarian development and local responses. Special issue land. https://www.mdpi.com/journal/land/special_issues/landgrabbing

Haller, T., & Weissman, S. (2023). The great 'Anti-politics' progress machine: Mega-infrastructure projects, disenchanted institutional change and dramas of grabbed commons. In A. Neef et al. (Eds.), *Routledge handbook of global land and resource grabbing* (pp. 358–372). Routledge, Taylor & Francis.

Haller, T., & Weissman, S. (Eds.). (2024). *Disenchanted modernities. Mega-infrastructure projects, socio-ecological changes and local responses.* Lit.

Haller, T., et al. (2018). *Paradigm change or old wine in new bottles? Debating and reformulating SDGs—An experiment.* University of Bern. http://www.anthro.unibe.ch/unibe/portal/fak_historisch/dkk/anthro/content/e40416/e96353/e96354/files747906/SDG_Text_FInal_ger.pdf

Hardin, G. (1968). The tragedy of the commons. *Science*, New Series, 162(3859), 1243–1248. http://www.jstor.org/stable/1724745.

Helfrich, S., & Bollier, D. (2015). *Die Welt der Commons: Muster gemeinsamen Handelns* (p. 384). Transcript Verlag.

Joint SDG Fund. (2022). Innovative mechanism of the United Nations. https://www.jointsdgfund.org/sdg-financing#:~:text=The%20Joint%20SDG%20Fund%20is,clear%20path%20to%20self%2Dsufficiency

Kronenburg Garcia, A. K., Haller, T., van Dijk, H., Samimi, C., & Warner, J. (Eds.). (2022). Drylands facing change: Interventions, investments and identities.: Routledge, Taylor & Francis. Open access: https://www.taylorfrancis.com/books/oa-edit/10.4324/9781003174486/drylands-facing-change-angela-kronenburg-garc%C3%ADa-tobias-haller-cyrus-samimi-han-van-dijk-jeroen-warner

Larsen, P. B., Haller, T., & Kothari, A. (2022). Sanctioning disciplined grabs (SDGs): From SDGs as green anti-politics machine to radical alternatives? *Geoforum, 131*, 20–26.

Netting, R. McC. (1981). Balancing on an alp: Ecological change and continuity in a Swiss mountain community.: Cambridge University Press.

Neumann, R. P. (1998). *Imposing wilderness: Struggles over livelihood and nature preservation in Africa.* University of California Press.

Ostrom, E. (1990). *Governing the commons.* Cambridge University Press.

Quintana, A., & Campbell, L. M. (2019). Critical commons scholarship: A typology. *International Journal of the Commons, 13*(2), 1112–1127.

Ramutsindela, M., Matose, F., & Mushonga, T. (2022). *The violence of conservation in Africa: State, militarization and alternatives*. Edward Elgar.

Sarker, A., & Blomquist, W. (2019). Addressing misperceptions of governing the commons. *Journal of Institutional Economics, 15*(2), 281–301.

Schläppi, D. (2018). Einleitung. In D. Schläppi & M.-C. Gruber (Eds.), *Von der Allmende zur Share Economy. Gemeinbesitz und kollektive Ressourcen in historischer und rechtlicher Perspektive* (pp. 9–70). Berliner Wissenschaftsverlag.

Soliev, I., & Theesfeld, I. (2020). Benefit sharing for solving transboundary commons dilemma in Central Asia. *International Journal of the Commons, 14*(1), 61–77.

WCED. (1987). Our common future. Report of the World Commission on Environment and Development.

Weissman, S. (2019). Discourse and entanglement in a transnational conservation arena: Deciphering the ideologies and narratives behind conservation discourse in the 'glocal' commons in Kenya. In T. Haller, T. Breu, T. de Moor, C. Rohr, & H. Znoj (Eds.), *The commons in a Glocal world: Global connections and local responses* (pp. 392–413). Routledge.

Werthmüller, F. (2024). Fishing in troubled waters: The impacts of LAPSSET on the local fisheries in Lamu, Kenya. In Haller, T, & Weissman, S. (Eds.). in press. Disenchanted modernities: Mega-infrastructure projects, socio-ecological changes and local responses.

CHAPTER 15

Epilogue: The Sámi of the Kola Peninsula and Industrial Companies

Andrei Danilov

That humans are part of, an extension of, nature and our Earth is obvious to Indigenous peoples in the world. Nature and Indigenous peoples are an inseparable whole. The damage that contemporary industrial and extractive lifestyles cause to nature hurts us, Indigenous people, and is an important concern for us. Since ancient times, every aspect of Indigenous peoples' lives and cultures has been grounded in unity with the natural environment. And we, Indigenous people, strive to preserve the land and the wisdom that we inherited from our ancestors and pass it on to future generations. The exploitation of our territories by industrial companies destroys our lands and causes enormous harm to the climate.

The Kolmozerskoye lithium deposit in Murmansk Region is planned to be exploited in the near future. Two of Russia's largest industrial companies, Nornickel[1] and the State Corporation Rosatom,[2] have created a new daughter company, Polar Lithium, to conduct the work. This deposit is located on current grazing lands where reindeer herding has been carried out as an uninterrupted traditional economy since pre-Soviet times. The company and the regional authorities have been generously reporting in public media about the promises of the new mining project for the economic and social development of the region, where new resources will generate profit and more employment opportunities; for the sustainable energy transition, where Russia can maintain

[1] https://nornickel.com/
[2] https://www.rosatom.ru/en/index.html

A. Danilov (✉)
Sami Heritage and Development Fund, Melbu, Norway

R. Toivanen et al. (eds.), *Decolonizing the Sustainable Development Goals*,
https://doi.org/10.1007/978-3-032-14263-4_15

its position as a major producer of minerals for increased battery production; for social justice, where Indigenous people have been invited in the planning stage of the new mine, and its negative impact on traditional economy and culture is transparently discussed and fairly compensated. Such reports conceal the reality that the affected territory will be transformed in a way that will make it unavailable, lost forever for reindeer herding and fishing for my Sámi people.

Lithium is also a metal that is needed, among other things, for the Russian defense industry. This need brings an extra emergency in the development of this deposit and is used as a justification by state authorities to accelerate all procedures that enable the deposit's exploitation. After local Sámi people expressed some concerns and negative comments in the media, Polar Lithium quickly put an announcement for a so-called consultations (in Russian) meeting with the local Indigenous community in Lovozero District. It has created a website where it claims that it has attracted international experts in the evaluation of the project's impact on Indigenous culture and economy.

Many Sámi people, however, see this consultation process mainly as an attempt of the company to show that "a dialogue" has been initiated, and Polar Lithium adheres to the principle of Free, Prior, and Informed Consent. In reality, during these "consultations," deliberately announced at a short notice and attended by only a small number of Indigenous people, the company treated the representatives of the local population as an "audience" and simply informed them about the planned activities that would cause irreversible destruction of reindeer grazing lands. This is a dialogue where the Indigenous side is treated "with a gun to the temple" (Fig. 15.1).

Let me give you an example: on June 5, 2022, an event was held in the village of Lovozero in the form of "consultations with the Sámi and the local population." Galina Shebut, head of the village of Lovozero, spoke: "Perhaps the result to it [the consultation] will be of interest to patriotically minded residents of our country, and I think that most of us are patriots nowadays. I am an ordinary resident of the country, like most of you. Is this production, which has been initiated, important for substituting international imports in connection with current events? The population should understand how important this development is at the current moment."

Let me clarify that "the current moment (or current events)" is a fixed phrase in Russia for the war in Ukraine. Thus, we see how the authorities directly state during the "consultations" that this production is important for Russia in the context of the war on Ukraine. Nevertheless, at this meeting, two leaders of the Sámi movement, Alexander Slupachik and Andrey Zhvavy, spoke up and voiced Sámi people's concerns about and negative attitude to the project. Both were subsequently forced to leave Russia and are now amidst an unpredictably long and insecure process of seeking political asylum in Norway.[3]

The second example of discrimination against Indigenous people and interests in the Arctic in the name of developing green technology is the mining

[3] https://kolmozer.ru

Fig. 15.1 The shores of Kol'mozero Lake, where Reindeer Herding Team 8 used to live and graze their reindeer. (In 2021, long before the Kolmozerskoye lithium deposit received approval for exploitation, the Reindeer Herd 8 territory was taken over and filled with temporary accommodations for miners exploring the deposit. Photo: Vladislava Vladimirova, June 2012)

project of the platinum deposit in the Fedorova Tundra by the company Fedorovo Resources (https://fedorovoresources.ru/ru/). The company has already stated that traditional economic practices have not been carried out on these lands, which is untrue. The company states in its plan that an assessment of the impact on the social and environment will be carried out in accordance with international standards and the requirements of the International Finance Group and the World Bank—environmental and social impact assessment (ESIA). However, the Sámi have not been properly involved in this process. Given my previous experience of Russian and international mining in the Russian North, I doubt that the voice of the Sámi will be heard or have any impact. My people, the Kola Sámi, have repeatedly stated that they are against this project and pointed to its destructive impacts on Sámi culture and local people's land use and relations to nature.

As a representative of the Sámi people, a member of the Sámi community in the village of Lovozero, and a head of the Sámi Heritage and Development Fund, I filed an application for the formation of a territory of traditional land use within the Fedorova Tundra, an Indigenous right provided in Federal Law.

However, neither the regional nor the federal authorities bothered to give any specific answer to my application. They refused to satisfy this legally sound request, inventing various bureaucratic pretexts. Then I contacted the Fedorovo Resources company to ask them for help organizing a territory of traditional land use in Fedorova Tundra. But to date, my proposal has remained unheard. Thus, the Sámi will be deprived not only of their land but also of their legal right to lead a traditional way of life. The company's website has the latest press release from July 7, 2022. Thus, we lack any newer updates about whether the project is progressing or is on hold. The company does not seem to conduct any visible activities.

The third example is the transformation of the Seidozero Nature Reserve into a regional nature park. On the initiative of the Sámi, this nature conservation site was created in 2003 since Seidozero Lake is a sacred place for the Sámi. This is a place of great historical and religious significance for us, where Sámi have been conducting their ritual practices, and our sacred heritage is located. In recent years, bureaucratic procedures and problems, such as online purchase of visit tickets and subsequent failures of the online registration system, have led to multiple cases of Sámi not being allowed to visit sacred places. At the moment, changes have been made to Murmansk Region law on protected natural territories, and the reserve has been transformed into a nature park. The priority goals of this new organizational form are support for mining and tourism development! The Sámi are allocated a small area of territory which is practically unsuitable for reindeer herding. In addition, in order to use it, the formal permission of the new park director is required. The Sámi, local residents, and non-profit organizations wrote appeals to the President of Russia and to regional authorities, where they express their sadness, disappointment, and anger about this transformation. Further, at a meeting with local residents, the Sámi spoke out against it. So far, no response has been sent to these letters and expressions of concern and objection.

How are industrial development projects coordinated with Indigenous peoples in Russia? I know from my own experience how the principle of obtaining Free, Prior and Informed Consent from Indigenous peoples is (not) observed. Companies employ what they call "international experts" who in reality are clerks who are loyal to government bodies and industrial companies. Companies and experts describe politically correct procedures and results that accord with international standards but their words diverge hugely from their deeds. So, I was personally involved in the development of Nornickel's policy toward Indigenous peoples. But at the time when the consultations were held and the policies discussed, they had already been formally adopted by the company, so the consultations could not have any impact. What is this if not a deception and a farce on the part of Nornickel and their "international experts"!

Many more examples of land grabbing, discrimination, and disrespect can be presented here. But they cannot stop us from struggling to be able to use our lands, to have a future, and to preserve ourselves as a people. One of the successful examples is my lawsuit for preserving the right to traditional hunting as a way of Indigenous subsistence and life. Three long years of legal action finally resulted in the satisfaction of my claim in the Constitutional Court of Russia.

Yes, we are persecuted, and we have to face the unequal forces of powerful business actors like mining companies and same-minded state bureaucratic authorities. That is why I, as an activist and human rights defender, was forced to flee from Russia in 2022 after my administrative arrest under the pretext of resisting a police officer. It was a very difficult decision and a personal drama! But I did not leave the land of my ancestors; I came to one of the Sápmi countries (Norway), which is part of the land of the cross-border Sámi nation. And Sámi people support me at all levels.

I want to end my story with the words from the film *Kautokeino rebellion* (*Kautokeino-opprøret*): "Some things are hard to talk about, but silence leads to nothing. We must talk and try to understand even when there is no hope. Hope does not die; it moves to new hearts. Like the spring wind that brings the scent of summer."

Correction to: Decolonizing the Sustainable Development Goals

Reetta Toivanen, Vladislava Vladimirova, and Carl-Gösta Ojala

Correction to:
R. Toivanen et al. (eds.), *Decolonizing the Sustainable Development Goals*, https://doi.org/10.1007/978-3-032-14263-4

The book was inadvertently published with the missing funding information. This has now been corrected and approved.

The updated version of this book can be found at
https://doi.org/10.1007/978-3-032-14263-4

R. Toivanen et al. (eds.), *Decolonizing the Sustainable Development Goals*,
https://doi.org/10.1007/978-3-032-14263-4_16

Index[1]

[1] Note: Page numbers followed by 'n' refer to notes.

R. Toivanen et al. (eds.), *Decolonizing the Sustainable Development Goals*, https://doi.org/10.1007/978-3-032-14263-4

Kolibri 360 GmbH
Ferdinand-Jühlke-Straße 7
99095 Erfurt, Deutschland
produktsicherheit@kolibri360.de